Springer Theses

Recognizing Outstanding Ph.D. Research

Aims and Scope

The series "Springer Theses" brings together a selection of the very best Ph.D. theses from around the world and across the physical sciences. Nominated and endorsed by two recognized specialists, each published volume has been selected for its scientific excellence and the high impact of its contents for the pertinent field of research. For greater accessibility to non-specialists, the published versions include an extended introduction, as well as a foreword by the student's supervisor explaining the special relevance of the work for the field. As a whole, the series will provide a valuable resource both for newcomers to the research fields described, and for other scientists seeking detailed background information on special questions. Finally, it provides an accredited documentation of the valuable contributions made by today's younger generation of scientists.

Theses are accepted into the series by invited nomination only and must fulfill all of the following criteria

- They must be written in good English.
- The topic should fall within the confines of Chemistry, Physics, Earth Sciences, Engineering and related interdisciplinary fields such as Materials, Nanoscience, Chemical Engineering, Complex Systems and Biophysics.
- The work reported in the thesis must represent a significant scientific advance.
- If the thesis includes previously published material, permission to reproduce this must be gained from the respective copyright holder.
- They must have been examined and passed during the 12 months prior to nomination.
- Each thesis should include a foreword by the supervisor outlining the significance of its content.
- The theses should have a clearly defined structure including an introduction accessible to scientists not expert in that particular field.

More information about this series at http://www.springer.com/series/8790

Aída Serrano Rubio

Modified Au-Based Nanomaterials Studied by Surface Plasmon Resonance Spectroscopy

Doctoral Thesis accepted by
the Universidad Complutense de Madrid, Spain

Springer

Author
Dr. Aída Serrano Rubio
Departamento de Electrocerámica
Instituto de Ceramica y Vidrio
Madrid
Spain

Supervisors
Prof. Óscar Rodríguez de la Fuente
Departamento de Física de Materiales
Universidad Complutense de Madrid
Madrid
Spain

Dr. Miguel Ángel García
Departamento de Electrocerámica
Universidad Complutense de Madrid
Madrid
Spain

ISSN 2190-5053 ISSN 2190-5061 (electronic)
Springer Theses
ISBN 978-3-319-37016-3 ISBN 978-3-319-19402-8 (eBook)
DOI 10.1007/978-3-319-19402-8

Springer Cham Heidelberg New York Dordrecht London

Softcover reprint of the hardcover 1st edition 2015

Printed on acid-free paper

Springer International Publishing AG Switzerland is part of Springer Science+Business Media
(www.springer.com)

Education is the most powerful weapon which you can use to change the world

Nelson Mandela, 1918–2013

Experimental physics cannot do without instruments

Jean-Antoine Nollet, 1700–1770

To my parents and my siblings
To Raúl

Supervisors' Foreword

Surface plasmon resonance (SPR) is, unquestionably, the most interesting optical property of metallic nanostructures. It consists of a collective oscillation of conduction electrons close to the surface raising a local charge accumulation at the metal/dielectric interface. For some metals (mainly noble metals) this oscillation is very weakly damped. As a consequence, the SPR creates local electromagnetic fields that can be up to two orders of magnitude larger than those required to excite it. Thus, SPR allows tailoring optical properties at the nanoscale in a way that is not possible by using dielectric optics elements. SPR overcomes the diffraction limit, opening the possibility to develop new optical circuits with very significant advantages with respect to electronic ones. The very weak damping makes also SPR extremely sensitive to any modification of the local environment of the interface. Therefore, SPR can be used for sensing purposes, as it is able to detect the deposition of sub-monolayer films or small variation in their physical properties. Consequently, the development of plasmonics has been extremely fast in the past years, and nowadays SPR is exploded in many fields such as biomedicine, energy harvesting, environmental protection, or information technologies.

In addition to the inherent properties of plasmonic systems, further advantages can be gained by combining surface plasmons with other systems in hybrid nanostructures. Besides exploding the properties of each type of particle, their coexistence may induce interaction effects resulting in an improved behavior. Hence, there is a growing interest in hybrid nanostructures containing a plasmonic element with other functional systems. A key milestone for the advance of this field is the fabrication of those hybrid nanostructures. Recent advances in colloidal chemistry provide excellent results in terms of control of morphology and size distributions. However, the amount obtained in each synthesis is certainly small (of the order of milligrams) and results quite expensive for applications requiring a large number of nanoparticles. Therefore, the development of new processing methods is still an open issue.

In her Ph.D. thesis, Aída Serrano has addressed the coupling of SPR with other physical properties in hybrid nanostructures. For this purpose, she has developed a new method to fabricate hybrid noble metal-transition metal oxide nanostructures.

The method, consisting in the deposition of multilayers and subsequent thermal treatments under a controlled atmosphere, allows covering large substrates with nanoparticles, with morphologies which can be controlled through the processing parameters. In addition, the method is easy to implement and economically competitive. Aída Serrano has performed a systematic study for the case of Au and Fe oxide nanostructures, but the system can be used for other combinations of metals and oxides.

Another interesting issue is the simultaneous combination of SPR with other types of radiation in order to explore the interaction of light with matter, taking advantage of the above-mentioned extreme sensitivity of SPR to local modifications in the region close to the interface. Regarding this, Aída Serrano has carried out a study on the effect of X-rays on different types of materials using SPR as a probe. For this purpose a novel SPR compatible with a synchrotron beamline has been developed. This device allows combining SPR with X-ray Absorption Spectroscopy (XAS) in situ and in real time. Using this device, Aída Serrano has studied the effects of X-rays in glasses, organic molecules, and transition metal oxides. Thanks to the capabilities of the device, it has been possible to identify the transient effect and the dynamics of the interactions between matter and light.

In summary, Aída Serrano's thesis has included the development of new processing methods and new instrumentation, as well as the study of new interaction effects between SPR and other systems, which has allowed her to achieve a high degree of qualification in scientific research.

Madrid, Spain
March 2015

Prof. Óscar Rodríguez de la Fuente
Dr. Miguel Ángel García

Preface

In this work, surface plasmon resonance (SPR) spectroscopy has been employed as a probe for studying modifications in nanostructured systems based on Au. These modifications include: (a) structural and morphological changes induced by annealing and (b) electronic modifications upon X-ray irradiation. The use of SPR spectroscopy to follow the evolution of modifications in nanostructures (especially, in situ and in real time) has been scarcely explored and its study results interesting and relevant for the plasmonic technology. For this, different experiments have been planned and performed:

- Fabrication of nanostructures from Au films and Au/Fe bilayers annealed under different conditions. Study by SPR spectroscopy of samples in order to follow the morphological changes of the systems varying the sample features (i.e., initial thickness) and the annealing conditions.
- The design and development of a SPR system based on the Kretschmann-Raether configuration compatible with an X-ray absorption spectroscopy (XAS) beamline (SPR-XAS setup).
- Study, using the SPR-XAS device, of the effects of the X-rays on glasses and Co-phthalocyanines (CoPcs). For the study on glasses, two different types are analyzed: soda-lime and silica substrates, and for the case of the CoPcs, these are grown varying the film thickness and the growth conditions.

For the realization of this work, samples have been fabricated by thermal evaporation and the characterization has been carried out using diverse spectroscopies and microscopies. For the case of studying surface plasmons (SPs), optical absorption spectroscopy and an experimental setup in the Kretschmann-Raether configuration, designed and developed for this work, have been employed.

In the first set of experiments, we have used SPR spectroscopy to study structural modifications of Au films and Au/Fe bilayers annealed under different conditions. For the Au films, these have been deposited on soda-lime substrates, varying their thickness, and have been further annealed in air at different temperatures, from 300 up to 500 °C. Films without annealing exhibit extended surface plasmons (ESPs), which depend on the film thickness [1, 2]. When samples are annealed, the thermal treatments promote modification of the continuous films toward Au isolated islands, which show morphological features that depend on the Au film initial thickness and the annealing temperature [3–5]. In this work, we follow the transition from ESPs in the initial films toward localized surface plasmons (LSPs) in the final Au nanostructures. Besides, features of SPR are correlated with the structural and morphological modifications induced in the films with the annealing. Specifically, for films with initial thickness below 30 nm, upon annealing, the Au NPs exhibit LSPs while thicker films lead to large islands (dimensions larger than visible light wavelength) that can exhibit ESPs [6].

On the other hand, for the case of Au/Fe bilayers, these are grown on silica substrates under various configurations and annealed in air or low vacuum at different conditions of the thermal treatments. Bilayer annealing represents a new method to fabricate large areas of complex nano and microparticles, which we present here for the first time. The morphological features of the complex particles depend on parameters such as layer initial thicknesses, the disposition of the layers, and the annealing conditions (i.e., temperature, atmosphere) [3, 7]. In this case, annealing also induces the Fe oxidation [8]. Thus, depending on the annealing atmosphere we obtain Au/α -Fe_2O_3 or Au/α -Fe_2O_3− γ -Fe_2O_3 nanostructures. In this way, we can promote the separation of Au and FeO_x nanostructures or the growth of complex dimers using the proper annealing conditions. SPR is modified depending on the different strategies carried out to fabricate the complex structures, since both Au and FeO_x particles exhibit different morphological features that affect the SPs.

In order to study changes induced by an external source in matter using SPs as probe thanks to their high sensitivity, we choose the XAS technique to combine it with the SPR, since it is well known that X-rays induce modifications in many kinds of materials [9–13]. The combination of both techniques, to our knowledge, has never been achieved. Thus, in this work, we have designed and developed a setup combining SPR and XAS: the SPR-XAS setup, which allows different types of measurements in situ and in real time, recording SPR or XAS signals while scanning other parameters such as time, energy, etc. This device is currently available for future experiments at the ESRF beamline BM25A-SpLine [14].

With this SPR-XAS setup, SPR spectroscopy has been used to study the changes in the refractive index induced by X-rays on glasses and CoPcs. For the case of glasses, Au films are grown on glass substrates using two types of glass substrates: soda-lime and silica substrates. In soda-lime glasses, X-ray irradiation induces color centers: HC1 and HC2 centers [15, 16]. Comparing the experimental results with simulated data, we observe that both the real and the imaginary parts of the refractive index of soda-lime glasses change upon irradiation in time intervals of a few minutes. After X-ray irradiation, the effects are partially reversible and the defects responsible for these modifications fade by recombination with electrons. The kinetics of the defect formation and fading process are also studied in situ and real time, obtaining a slow process with a characteristic time of the order of months and a fast process with a characteristic time of the order of minutes, which has not been previously reported [17]. This component has been measured thanks to the device developed, which allows monitoring changes during and immediately after irradiation. For the case of silica glasses, defects that modify their refractive index have been identified, but they are not detected at the laser wavelength used in the device. However, slight variations, of the order of 0.01 %, are observed and attributed to changes where the Au thin film is involved.

With respect to the study of CoPcs irradiated with hard X-rays using SPR as a probe, CoPc/Au samples are grown on silica substrates varying the CoPc thickness (from 2 to 10 nm) and the growth temperature: room temperature (RT) and 200 °C [18]. X-ray irradiation generates modifications on SPR spectra depending on CoPc thickness and growth temperature. The changes in SPR reflectivity and the full width at half maximum (FWHM) are larger for samples grown at RT than for those grown at 200 °C and independently of the growth temperature, thicker CoPc films show larger FWHM modifications of SPR spectra. Monitoring the reflectivity as a function of time, in real time and in situ, we observe that the induced reflectivity changes on samples are accumulative with time and slowly recover. Comparing the experimental results of SPR spectra with simulated data, we observe that the refractive index of the CoPc layers changes around 1.5–5 % depending on their thickness and growth conditions. These modifications of the refractive index are not related to structural changes and they are associated with electronic modifications induced by the irradiation [13].

The main conclusions related to this work are:

- The annealing of Au films or Au/Fe bilayers grown on glass substrates provides a method to modulate the morphological properties of the samples and with that, a process to tune the plasmonic properties in a wide range of applications and over large areas not easily achievable by other methods. Specifically, annealing of Au/Fe bilayers allows fabricating complex nanostructures combining plasmonic properties with others as catalytic or magnetic ones.
- Development of an experimental setup to study the effect of X-rays in real time and in situ. Useful information about the refractive index is obtained comparing the experimental results of SPR with simulated spectra: about the formation and fading processes of color centers in glasses and about the electronic changes varying parameters such as the thickness or the growth temperature of CoPc layers.

References

1. M.L. Brongersma, P.G. Kik, *Surface Plasmon Nanophotonics* (Springer, Berlin, 1988)
2. H. Raether, *Surface Plasmons on Smooth and Rouge Surfaces and on Gratings. Nanophotonics* (Springer, Berlin, 1988)
3. A.E.B. Presland, G.L. Price, D.L. Trimm, Hillock formation by surface diffusion on thin silver films. Surf. Sci. **29**(2), 424–434 (1972)
4. J.A. Thornton, D.W. Hoffman, Stress-related effects in thin films. Thin Solid Films **171**(1), 5–31 (1989)

5. I. Doron-Mor, Z. Barkay, N. Filip-Granit, A. Vaskevich, I. Rubinstein, Ultrathin gold island films on silanized glass. Morphology and optical properties. Chem. Mater. **16**(18), 3476–3483 (2004)
6. A. Serrano, O. Rodríguez de la Fuente, M.A. García, Extended and localized surface plasmons in annealed Au films on glass substrates. J. Appl. Phys. **108**(7), 074303 (2010)
7. C.V. Thompson, Solid-state dewetting of thin films. Annu. Rev. Mater. Res. **42**(1), 399–434 (2012)
8. H. Nelson, The low temperature oxidation of iron. J. Chem. Phys. **6**(10), 606–611 (1938)
9. R. Livingston, C.E. Nurnberger, A kinetic interpretation of the coloring of glass by x-rays. J. Phys. Chem. **39**(7), 1011–1020 (1934)
10. E. Itälä, *Fragmentation processes in organic molecules induced by synchrotron radiation*. PhD thesis, University of Turku, Finland, 2012.
11. E. Friebele, *Radiation Effects in Optical and Properties of Glass* (Westerville, New York, 1991)
12. H.S. Nalwa (ed.) *Silicon-based Materials and Devices* (Academic Press, USA, 2001)
13. P.L. Cook, P.S. Johnson, X. Liu, A.L. Chin, F.J. Himpsel, Radiation damage in biomimetic dye molecules for solar cells. J. Chem. Phys. **131**(21), 214702 (2009)
14. A. Serrano, O. Rodríguez de la Fuente, V. Collado, J. Rubio-Zuazo, C. Monton, G. Castro, M. García, Simultaneous surface plasmon resonance and x-ray absorption spectroscopy. Rev. Sci. Instrum. **83**(8), 083101 (2012)
15. J. Sheng, K. Kadono, T. Yazawa, Fading behavior of X-ray induced color centers in soda-lime silicate glass. *Applied radiation and isotopes : including data, instrumentation and methods for use in agriculture, industry and medicine* **57**(6), 813–817 (2002)
16. J. Sheng, K. Kadono, Y. Utagawa, T. Yazawa, X-ray irradiation on the soda-lime container glass. Appl. Radiat. Isot. **56**(4), 621–626 (2002)
17. A. Serrano, F. Galvez, O. Rodríguez de la Fuente, M.A. Garcia, X-ray irradiation of soda-lime glasses studied in situ with surface plasmon resonance spectroscopy. J. Appl. Phys. **113**(11), 113104 (2013)
18. A. Serrano, C. Monton, I. Valmiansk, F. Gálvez, A.L. Cortajarena, J. de la Venta, I. K. Shuller, M.A. García, Study of Co-phthalocyanine films by surface plasmon resonance spectroscopy J. Appl. Phys. **115**, 103106 (2014)

Acknowledgments

This thesis would not have been possible without the contribution of many people, research centers and Universities. First, I would like to thank Dr. Miguel Ángel García and Dr. Óscar Rodríguez de la Fuente, my Ph.D. supervisors, for believing in me. I was lucky to have you both at work and at personal level. I would like to acknowledge Dr. Germán Castro for his strong support in the development of the SPR-XAS setup at the SpLine BM25 beamline at ESRF in Grenoble, France. I am very grateful to Prof. José Francisco Fernández, Prof. José Llopis, Prof. Juan Rojo, Prof. Iván Shuller and Dr. César de Julián Fernández for allowing me learning in their laboratories and for many useful discussions. Thanks also my colleagues at the Instituto de Ceramica y Vidrio, CSIC. Finally, I would like to thank my big family and Raúl for all of the support throughout my Ph.D.

Contents

Chapter 1
Introduction: Surface Plasmons

This chapter presents a review of surface plasmons in nanoparticles and thin films. The fundamentals of surface plasmons as well as their application fields are summarized. This chapter also includes the objectives of the work and an overview of the organization of the manuscript.

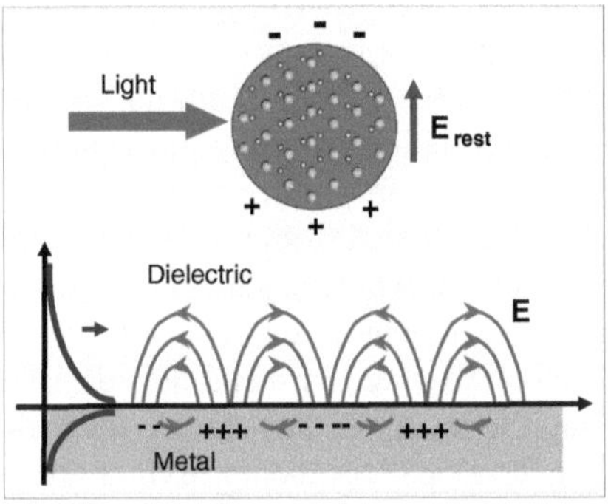

1.1 Introduction

Surface plasmons (SPs) are one of the most outstanding properties of metallic nanostructures. They consist of a collective oscillation of conduction electrons at the interface between a metal and a dielectric medium. While SPs can be excited at the surface of bulk metals, they are particularly important in the case of nanostructures [1–3]. The large surface to volume ratio of nanostructures allows that SPs determine their optical properties, providing new phenomena not achievable in bulk materials.

Depending on the geometry of the nanostructure, there are different types of SPs. In nanoparticles (NPs) with 0−dimension (i.e., the dimensions of the nanoparticles are much smaller than the wavelength of the incident light), SPs consist of charge density oscillations in phase within NPs. These plasmons are characterized by a

A. Serrano Rubio, *Modified Au-Based Nanomaterials Studied by Surface Plasmon Resonance Spectroscopy*, Springer Theses, DOI 10.1007/978-3-319-19402-8_1

resonant frequency and they are called localized surface plasmons (LSPs) [4]. For the case of thin films (with dimensions in plane much larger than wavelength of the incident light), SPs are propagating waves with certain dispersion relation and they are known as extended surface plasmons (ESPs) [5]. Finally, for mesoscopic systems (with dimensions similar to the light wavelength), SPs are stationary waves at the interface [2, 3]. The interference of the initial waves with those reflected at the edge of the nanostructures is, on average, destructive, except for certain wavelengths that are an integer number of times the nanostructure dimension. For these particular wavelengths, the interference yields stationary waves, known as plasmon modes [1–3].

We can make an analogy between SPs and mechanical oscillations: LSPs, ESPs and plasmon modes should correspond to a mass with a spring, an infinite rope and an oscillating string, respectively. Only LSPs and ESPs are studied in this thesis.

1.2 Brief History

Long since LSPs have been used by artists to generate brilliant colors in glass artefacts and artwork, where for example, the inclusion of metal NPs with different size or shape into the glass created a large variety of colors [2]. One of the most famous examples is the Lycurgus cup dating from the 4th century A.D., which shows a green color when it is illuminated from outside and observed with reflected light and shines in red when it is illuminated from inside in transmitting light conditions. Other examples can be found in window glasses of churches or cathedrals (see Fig. 1.1). However, the first documented scientific studies about LSPs are found in 1904, when the bright colors observed in metal doped glasses [6] were described by Maxwell Garnett using the theory of Paul Drude for metals, proposed in 1900. Later, in 1908 Gustav Mie developed his theory of light scattering by homogeneous spherical particles [7]. The expansion for spheroidal particles was published in 1912 by Richard Gans for Au NPs and nowadays, the Mie-Gans theory is currently being applied in the field of nanotechnology [8]. For the case of more complex NPs, in 1964 Howard DeVoe introduced the discrete dipole approximation (DDA) method [9].

On the other hand, the first documented studies about ESPs date from 1902, when Robert Woods observed anomalies in optical reflection measurements on metallic gratings [10]. Later, the anomalies observed by Woods were studied by Ugo Fano and these were associated with electromagnetic waves on the surface of the metallic gratings [11]. In the fifties (1956), David Pines theoretically described the features of energy loss experienced by fast electrons traveling through metals and attributed them to collective oscillations of free electrons in the metal, which were called plasmons [12]. In 1957, Rufus Ritchie proposed the concept of SPs in the context of electron energy losses in thin films, showing that plasmon modes can exist near the surface of metals. The experimental verification was provided 2 years later by Powell and Swan [13]. Later, in 1968, Ritchie observed Wood's anomalies and described them in terms of surface plasmon resonances excited on the gratings [14].

Fig. 1.1 **a** Lycurgus cup dating from the 4th century A.D. and **b** a glass window in Cathedral of León dating from the 13–14th century A.D.

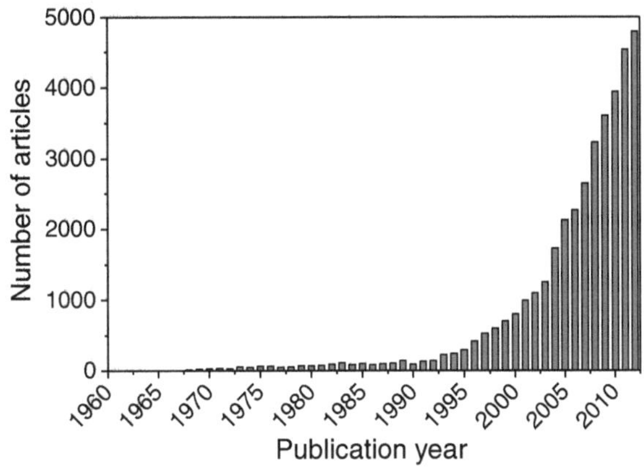

Fig. 1.2 Number of articles published annually containing the phrase "surface plasmon" in either title or abstract. Data provided from http://www.scopus.com

However, the major advance in the study of ESPs was made in 1968 when Otto [15], Kretschmann and Raether [16, 17] presented methods, easily accessible, for the optical excitation of SPs on metal films.

At that time, great advances in the SPs of metal nanostructures were achieved by authors as Kreibig and Zacharias [18] in 1970. In 1974, the term surface plasmon polariton (SPP) was introduced by Cunningham et al. [19] and the surface enhanced Raman scattering (SERS) was detected by Fleischmann et al. [20]. Since 1990, the growth of the field of plasmonics is clearly reflected in the bibliography, where an exponential growth is found, as Fig. 1.2 shows. Since then, the extended use of SERS, the possibility to tune the optical properties of metal nanostructures and the development of sensors based on SPs, are responsible for that quick growth into a wide range of applications [2].

1.3 Theory: Surface Plasmons

Most of the optical properties of bulk metals can be properly described using the Drude model, assuming that the metal is formed by a lattice of ionic cores and an ensemble conduction electron moving freely inside the metal. These conduction electrons can exhibit collective oscillations known as bulk plasmons, which in the framework of quantum mechanics are quantified. The quantum of the volume plasmon has an energy $\hbar\omega_p = \hbar\sqrt{\frac{4\pi ne^2}{m_0}}$ of the order of 10 eV, being ω_p the plasma frequency and n the electron density [5, 21].

An important extension of the plasmon physics has been achieved by the concept of SPs, where Maxwell theory, with the proper boundary conditions, shows that electromagnetic waves can propagate along a metallic surface with a broad spectrum of frequencies from $\omega = 0$ up to $\omega = \omega_p/\sqrt{2}$, depending on the wavevector k [1, 5]. The dispersion relationships of different types of plasmons are illustrated in Fig. 1.3. It is possible to observe that the dispersion relationship of SPs falls to the right of the light line with a larger wavevector than the light waves for same frequency values, propagating along the surface. Therefore, SPs are non radiative and they describe fluctuations of the surface electron density. SPs have their maximum intensity at the surface and their electromagnetic field decays exponentially in the direction perpendicular to the surface, which is characteristic of surface waves [5]. In the case of 0D systems (i.e., NPs), the electron oscillations cannot propagate and they do not exhibit a dispersion relationship but just a resonance frequency.

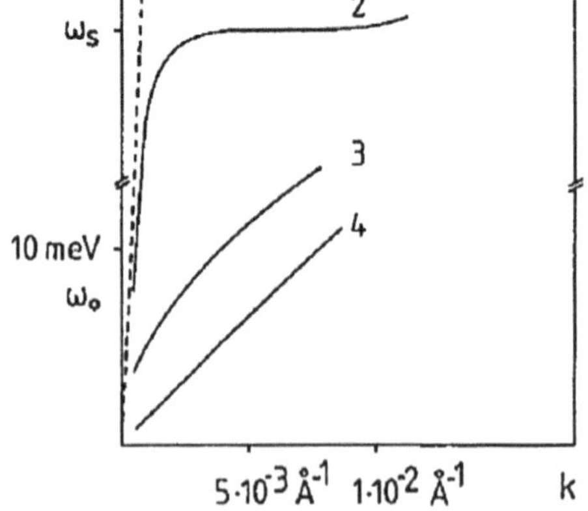

Fig. 1.3 Scheme of the dispersion relationships of different plasmons in a solid. *1* Volume plasmons, *2* surface plasmons, *3* plasmons in a 2D electron gas, *4* plasmons in a 1D system. The upper scale is valid for *1* and *2* and the scale below is valid for *3* and *4*. *ll* is the *light line*. Figure adapted from Raether et al. [5]

1.3.1 Surface Plasmons in Metal Nanostructures

For the case of metallic NPs with a size that is significantly smaller than the wavelength of the excitation light, the electromagnetic field of the light exerts a force on the conduction electrons moving them towards the NP surface, Fig. 1.4. Thereby, a negative charge is accumulated in one side and a positive charge in the opposite, creating an electric dipole. This dipole creates an electric field inside the NP opposite to that of the light, which acts as a restoring force to the equilibrium position. The situation is similar to that of the linear oscillator. If the electrons are displaced from the equilibrium position and later the field is removed, they oscillate with a frequency called resonance frequency (plasmonic frequency), which for most transition metals falls in the ultraviolet-visible (UV-Vis) region of the optical spectrum [2–4, 22].

In NPs, the electrons suffer damping during their movement due to the scattering process with the ionic cores and the NP surface, so the situation is similar to a damped oscillator, reducing the oscillation amplitude [2, 3].

Consequently, metallic NPs show absorption bands in the optical spectrum associated with the SPs. The NP absorption can be measured by the extinction cross section coefficient, σ_{ext}, which provides an estimation of the fraction of photons that are absorbed and scattered from the incident beam when it passes through the NPs [4]. Thus, σ_{ext} depends on both size and the absorption efficient of NPs. Figure 1.5 illustrates the wavelength of SP absorption band and the σ_{ext} for some metallic NPs with 10 nm size. The extinction cross section for a particle is expected to be its geometrical size at most. However, experimentally for metallic NPs at the SP resonant frequency, the extinction cross section can be up to 1000 times larger than the geometrical size of particles. The origin of these huge values for SP excitation in NPs is related to the nature of surface plasmon resonance (SPR). When SPR is excited, the light is not just absorbed to promote a transition from a ground state to an excited one (as in semiconductors or organic molecules). The electric field generated by the displacement of electrons interferes with that of the light, being at some points a destructive interference that yields light absorbtion. Since the field created by the displacement

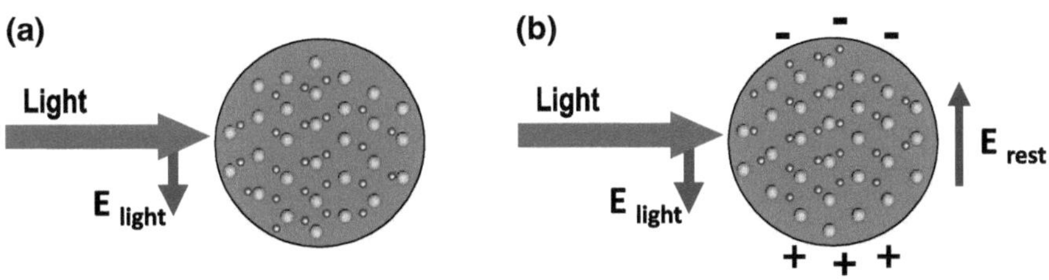

Fig. 1.4 **a** A metallic NP is illuminated with light and **b** when this NPs is illuminated the electric field displaces the conduction electrons yielding a charge accumulation at the NP surface. This charge creates a restoring field opposite to that of the light pushing the electrons towards the equilibrium position conditions

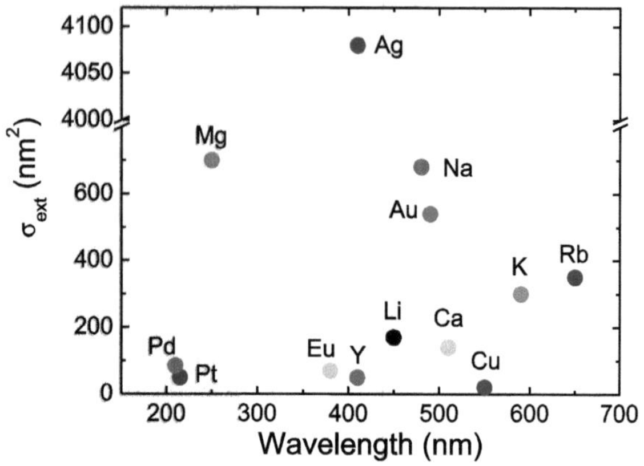

Fig. 1.5 Extinction cross section coefficient versus the wavelength of the SP absorption band for some metallic NPs with 10 nm size embedded in silica. Figure adapted from García [23]

of electrons is not limited to the NP volume but it spreads out, the effective absorption takes place in a volume much larger than the NP one [23].

The metallic NPs with higher extinction cross section coefficient are the Mg, Na, Ag and Au NPs. Specifically, the Mg and Na NPs are easily oxidized and thus, the Ag and Au NPs are the most interesting. However Ag is susceptible to sulphuration by atmospheric pollution or it can also be oxidized. Au shows appealing properties such as chemical stability, high biocompatibility and easy functionalization [1, 23]. Therefore, Au is the most frequently used element and the material employed for this thesis work.

The proper analysis of the SPs requires solving the Maxwell equations of electromagnetism in the metallic NPs with the proper boundary conditions for the surrounding medium [1, 7]. For that purpose, some theories and approximations are usually employed.

- **Mie theory**

Gustav Mie found the exact solution to the Maxwell equations for the optical response of spherical NPs of arbitrary size that are non-interacting, immersed in a homogeneous medium and illuminated with a plane monochromatic wave. The Mie theory describes the NP as a spherical object characterized by a dielectric function in an infinite dielectric medium [1, 7].

According to Drude model [24, 25], frequency dependent dielectric function can be described by the following expression:

$$\varepsilon(\omega) = 1 - \frac{w_p}{w^2 - i\omega\gamma} \tag{1.1}$$

where ω_p is the plasma frequency and γ is the damping constant due to the dispersion of the electrons. The damping constant is given by:

$$\gamma = \gamma_0 + a\frac{v_F}{R} \tag{1.2}$$

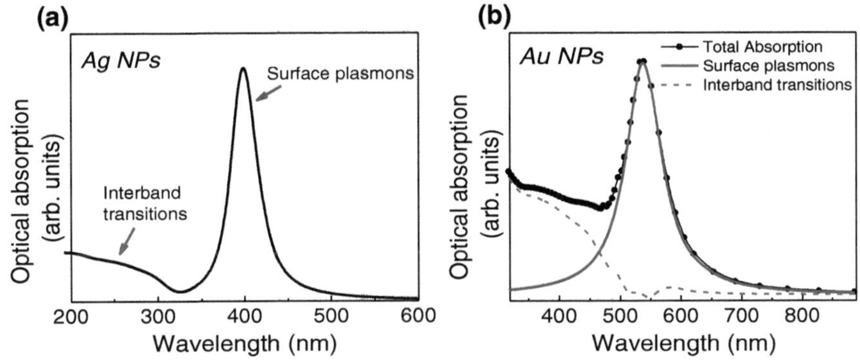

Fig. 1.6 Absorption spectra for **a** Ag and **b** Au NPs with 40 nm size (embedded in a silica matrix with $\varepsilon = 2.25$). For Ag NPs, the contribution of interband transitions and SPs are clearly resolved while for Au they overlap

where γ_0 represents the damping due to the scattering of the oscillating electrons with the ionic cores, a is a material dependent constant, v_F is the Fermi velocity and R is the NP radius. Therefore, including the damping term to the Drude model, the dielectric function can be expressed as:

$$\varepsilon(\omega, R) = 1 - \frac{w_p}{w^2 - i\omega(\gamma_0 + a\frac{v_F}{R})} \tag{1.3}$$

NPs can absorb energy through the following mechanisms: (i) collective excitations of the free electrons, SPs (ii) electron transitions of electrons from occupied to empty bulk bands, interband transitions and (iii) surface damping due to inelastic scattering processes [23, 26].

The optical spectrum for noble metal NPs presents the absorption ascribed to SPs plus an additional feature due to the interband transitions, ε_{inter} [23, 25]. Taking into account the contribution of the interband transitions, the dielectric function is given by the following expression:

$$\varepsilon(\omega, R) = 1 - \frac{w_p}{w^2 - i\omega(\gamma_0 + a\frac{v_F}{R})} + \varepsilon_{inter} \tag{1.4}$$

The absorption ascribed to interband transitions is clearly resolved in the case of Ag NPs since it falls at lower energies than SP band. However, in the case of Au NPs, it overlaps with the SP absorption band as Fig. 1.6 illustrates.

- **Discrete dipole approximation**

The discrete dipolar approximation provides equations to perform calculations on SPs in NPs assuming that they are spherical, isolated and their size is small compared with the wavelength of excitation light (besides the Mie theory). In this case, the electric field inside the NP can be considered as uniform most of the time and the particle can be described by an electric dipole [23, 26]. In this case, the solution of Maxwell equations becomes simplified and the extinction cross section coefficient is given by [1, 27]:

$$\sigma_{ext} = \frac{24\pi^2 R^3 \varepsilon_d^{3/2}}{\lambda} \frac{\varepsilon_2}{(\varepsilon_1 + 2\varepsilon_d)^2 + \varepsilon_2^2} \tag{1.5}$$

where R is the NP radius, λ is the light wavelength, ε_d is the dielectric constant of the surrounding medium and $\varepsilon_1 + i\varepsilon_2$ is the complex dielectric function of the metallic NPs, which depends on the wavelength.

1.3.1.1 Features of Surface Plasmons in Metal Nanostructures

Parameters such as the frequency (i.e., absorption maximum or color), bandwidth and the intensity of LSP absorption bands are characteristic of the type of material. Besides, these absorption bands are very sensitive to the size, distribution, shape, surrounding medium as well as interaction effects of metallic NPs [23, 26, 28].

- **Size effects**

The NP size affects on the SPR, distinguishing two size ranges: small NPs of less than 50 nm of diameter where the NP absorbs energy dominantly and larger NPs where the scattering effects dominate the response of NPs [23, 26].

For NPs smaller than 50 nm, when the size of homogeneous spherical nanostructures is fairly smaller than the wavelength of the incident light, the NPs feel the electromagnetic field and the displacement of the charges is mostly uniform. Thus, the NPs can be described by a dielectric dipole. In this case, only one resonance band is found. Parameters such as the width and the intensity of SPR band are modified by the NP size while the effect on the resonance frequency is quite reduced [1, 23, 26].

The size effect can be related to the damping of the electron oscillations. When SPs are excited, the electrons are damped in their movement by the scattering with the ionic cores and the surface. The relationship between the NP size and the full width at half maximum (FWHM), Γ, of the SPR band is given by:

$$\Gamma = a + \frac{b}{R} \tag{1.6}$$

where a and b are material dependent constants and R is the NP radius. As the NP size increases, the fraction of electrons in the shell close to the surface represents a smaller fraction of the oscillating electrons and thus, the total effective damping is reduced. The FWHM results inversely proportional to the NP radius.

Figure 1.7 shows the optical absorption spectra of Ag NPs embedded in silica (n = 1.5) with different sizes. The absorption band intensity increases with NP size, while the FWHM of the resonance peak decreases with the NP size. Figure 1.7b illustrates the law 1.6 for the Ag NPs with radius below 25 nm.

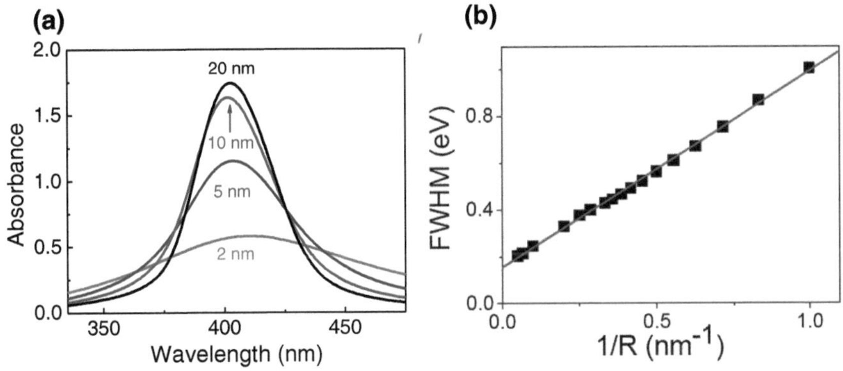

Fig. 1.7 **a** Optical absorption spectra for Ag NPs with different sizes in a medium with a refractive index n = 1.5 calculated according to the Mie theory and **b** linear relationship between FWHM and the inverse of the radius

For the case of NPs larger than 50 nm, these cannot be considered much smaller than the wavelength of incident light and the radiation effects are more important. In this case, the displacement of the electronic cloud is not homogeneous, even for spherical particles, and high multipolar charge distributions are induced [26]. NPs cannot be described as a dipole. Due to the radiation effects, the electrons lose energy experiencing a damping effect. In this way, the radiation damping reduces the intensity and makes SPR peaks broader and asymmetric [26]. Figure 1.8 shows the optical absorption spectra and the contribution of different terms (L = 1 dipolar term, L = 2 quadrupolar term and L = 3 octupole term, where L represents the terms of the multipolar expansion) for Au NPs with radius of 80 and 100 nm.

Despite the fact that the fabrication of NPs has been improved in the last years, there is always certain size dispersion. Due to this, the optical absorption spectrum of metallic NPs will correspond to an average of the optical absorptions of all NPs. As Fig. 1.9 shows, a broadening of the absorption band is induced by the size dispersion of NPs. Therefore, the NP size cannot be obtained just from the FWHM of the absorption band, since it depends on both average size and size distribution width.

- **Shape effects**

The SPR is strongly dependent on NP shape. Specifically, for the case of metal nanoellipsoids, there are three plasmon resonances corresponding to the oscillation of electrons along the three axes of the nanostructures. The resonance frequency depends on the orientation of the electric field relative to the NPs, therefore changing the axes length, the SPR of the nanostructures can be tuned [26]. A specific case can be the nanorods, which show two equal axes and where the charge accumulation at the NP surface will be different for electron oscillations along the rod axis (longitudinal plasmons) that in the perpendicular direction (transversal plasmons), as Fig. 1.10a illustrates. For nanorods, the SPR of transversal plasmons falls at similar position than for spherical NPs (at wavelengths slightly smaller than for spherical NPs) while

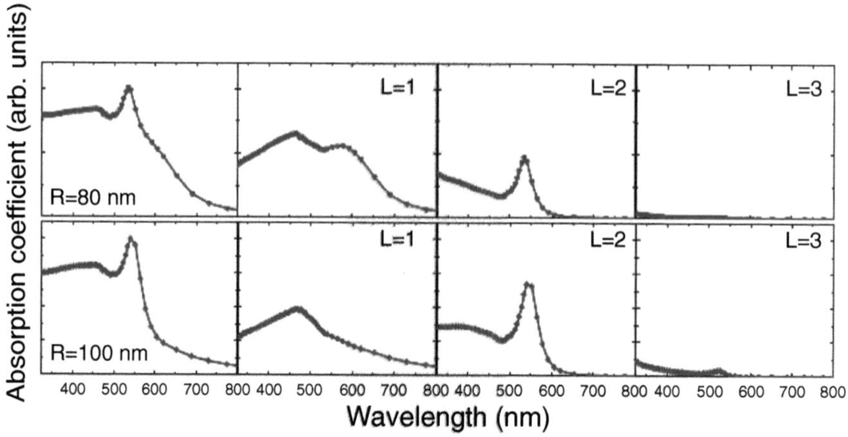

Fig. 1.8 Optical absorption spectra for Au NPs with radius of 80 and 100 nm in a medium with a refractive index n = 1.5 calculated according to the Mie theory and the contribution of the different multipolar terms

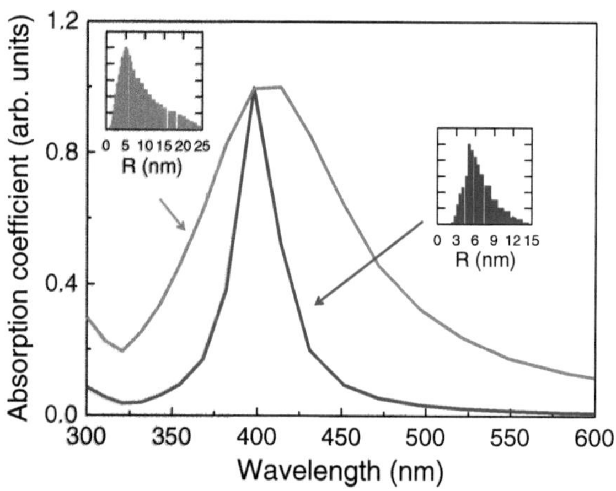

Fig. 1.9 Optical absorption spectra for Ag NPs ensembles with average 10 nm size with different size distribution in a medium with a refractive index n = 1.5 according to Mie theory. The size distributions of NPs are illustrated as *insets*

the resonance of longitudinal plasmons is shifted towards larger wavelengths when the aspect ratio of nanorods increases, as Fig. 1.10b shows. When the non-spherical NPs are macroscopically oriented in the same direction, it is possible to distinguish between the different resonances using polarized light [23, 26]. If the nanoellipsoids or nanorods are not oriented, an average of the different resonance modes is observed. NPs with other geometries, as triangular or cube shapes, show more complicated effects obtaining the plasmon resonance towards larger wavelength, if it is compared with that of the spherical NPs [23, 27].

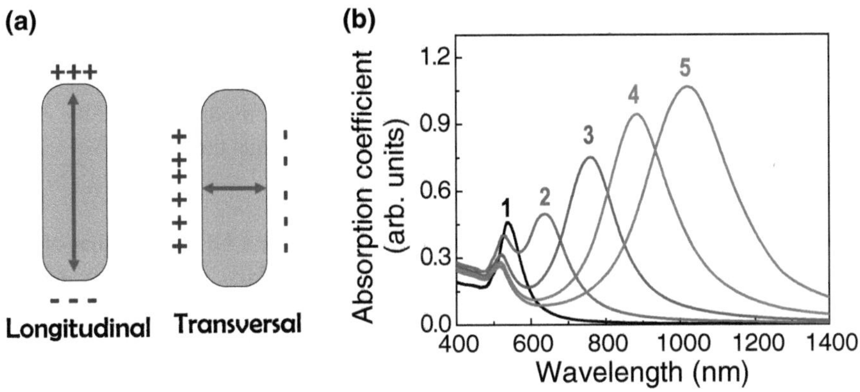

Fig. 1.10 **a** Scheme of charge accumulation region for longitudinal and transversal SPs and **b** calculated optical absorption spectra for Au nanorods with different aspect ratio indicated in the figure

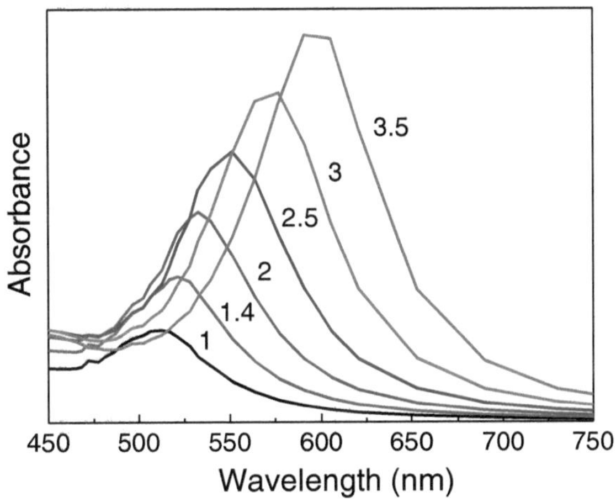

Fig. 1.11 Optical absorption spectra for Au NPs with 10 nm size in a medium with different dielectric function, indicated in the figure. Spectra are calculated according to the Mie theory

- **Medium effects**

The SPR depends on the surrounding medium besides the NPs features, shifting the plasmon resonance towards larger wavelengths when the dielectric function of the surrounding media increases, as Fig. 1.11 illustrates. The surrounding medium has a limited effect on the bandwidth that is mainly related to the NP size and the size distribution [4]. According to Eq. 1.5 (i.e., in the approximation dipolar for small NPs), the plasmon resonance intensity depends on both the dielectric function of the surrounding medium and the imaginary part of the dielectric function of the metal NPs, for the resonant condition $\varepsilon_1 = -2\varepsilon_d$. Specifically, for Au, the resonance band intensity increases with the dielectric function of the medium, as shown Fig. 1.11.

Other interesting case is when the physical environment of the NP changes, for example when NPs lie on a substrate or are capped with organic molecules to prevent agglomeration [25]. In this case, electromagnetic interactions of the bonds between NPs and molecules should be considered [1, 29], since the modification of the electronic configuration changes the electron oscillation and thus the SPR.

- **Interaction effects**

When disordered NPs are close enough, the local field at a NP will be that of the external incident light plus the field created by the rest of NPs. The field created by a NP decays distances of the order of nanometers. Thus, this field generated by a NP in the rest is not uniform, inducing non-uniform charge distributions inside the NP and yielding an extra damping of the electron oscillations, consequently. In general, interparticle interaction shifts the resonance band of SPs towards larger wavelengths and increases the FWHM [30, 31]. For the case of arranged NPs with particular geometries, it is possible to create controlled interferences of the electromagnetic field that allow additional tuning of the SPR [32].

1.3.2 Surface Plasmons in Metal Films

As previously indicated, for metallic films, SPs consist of oscillating waves of free electrons that travel at the interface between a metal and a dielectric. These electromagnetic waves, with p polarization or magnetic transversal, are associated with waves that travel perpendicularly to the metal/dielectric interface and decay exponentially into both media, as shown Fig. 1.12. The field amplitude of SPs is maximum at the metal/dielectric interface. SPs that couple to electromagnetic waves are called surface plasmon polaritons (SPPs) [5].

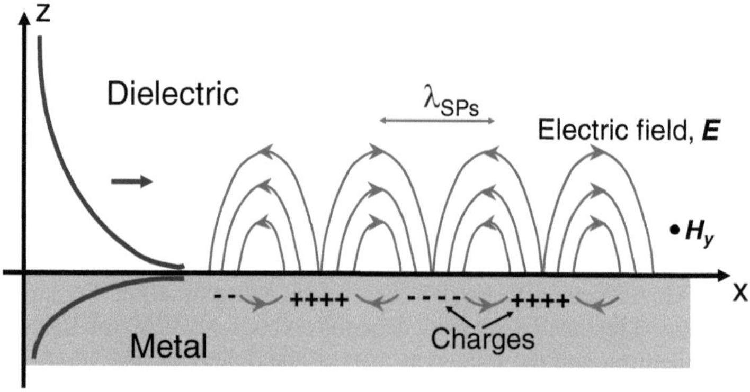

Fig. 1.12 SPs at the interface between a dielectric medium and a metal

1.3.2.1 Evanescent Waves

Evanescent waves are electromagnetic waves that travel along the interface between two media with different dielectric constants. These waves penetrate into the adjacent media decaying exponentially with the distance from the interface. SPs are evanescent waves that can be excited when a medium is a dielectric and the other one is a metal [33].

In order to understand the SPs, a mathematical description of evanescent waves is necessary [33]. An electromagnetic plane wave, that propagates in a medium with a certain refractive index, can be expressed by the electric field E if the exponent is complex [5]:

$$E = E_0 e^{(i\omega t - i\vec{k}\cdot\vec{r})} = E_0 e^{(i\omega t - ik_x x - ik_y y - ik_z z)} \quad (1.7)$$

where E_0 is the amplitude of the electric field, ω is the angular frequency, $\vec{k}$ is the wavevector, $\vec{r} \equiv (x, y, z)$ is the position vector and $i = \sqrt{-1}$.

The propagation direction of the wavevector is parallel to the propagation of the electromagnetic wave, thus:

$$k^2 = k_x^2 + k_y^2 + k_z^2 = (n\frac{2\pi}{\lambda})^2 = (n\frac{\omega}{c})^2 = \frac{\omega^2}{c^2}\varepsilon\mu \quad (1.8)$$

where λ is the wavelength, c is the light propagation velocity in vacuum, ε is dielectric constant of the medium and μ is the magnetic permeability.

Considering the electromagnetic wave is refracted at the interface between two media with refractive indexes n_1 and n_2 (see Fig. 1.13) and the system is two dimensional with $k_y = 0$, using the Snell's law $n_1 \sin\theta = n_2 \sin\beta$ ($k_{x1} = k_{x2} = k_x$) [33] and Eq. 1.8, the component of the wavevector k_{z2} perpendicular to the interface is given by:

$$k_{z2}^2 = n_1^2 \left(\frac{2\pi}{\lambda}\right)^2 \left(\frac{n_2^2}{n_1^2} - \sin^2\theta\right) \quad (1.9)$$

assuming than $n_1 \sin\theta > n_2 \sin\beta$. From Eq. 1.9, when $\sin\theta > n_2/n_1$, k_{z2} is purely imaginary.

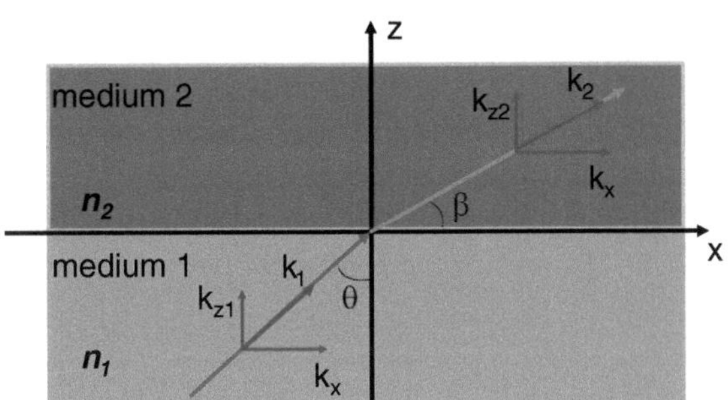

Fig. 1.13 Refraction of light at an incident angle θ, at the interface between two media with refractive indexes n_1 and n_2

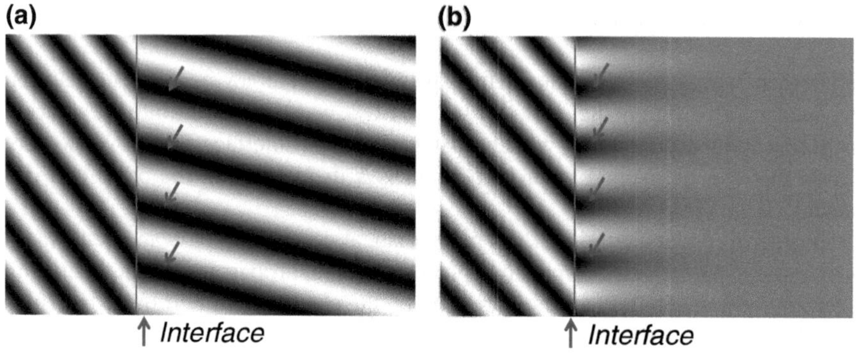

Fig. 1.14 A scheme of **a** a refracted incident wave and **b** an evanescent wave at an interface. They are marked with *red arrows* in the figure

Taking into account Eqs. 1.7 and 1.9, in medium 2, only longitudinal waves (i.e., waves with p polarization) traveling perpendicularly and decaying exponentially to the interface are obtained. This field is called evanescent field (Fig. 1.14) [5, 33].

$$E_2 = E_0 e^{-ik_z} e^{i(\omega t - k_x x)} \tag{1.10}$$

1.3.2.2 Dispersion Relation of Surface Plasmons

The relationship between the angular frequency, ω, and the wavevector, k, is known as the dispersion relation. For the interface between two media (see Fig. 1.14), a set of solutions of evanescent waves can be obtained. Using Eqs. 1.9 and 1.10, evanescent waves travel in the medium 1 (dielectric medium) and medium 2 (metallic medium) according to the following expressions [5, 34]:

For medium 1:

$$E_1 = E_0 e^{(k_x^2 - \varepsilon_1 \frac{\omega^2}{c^2})^{1/2} z} e^{i(\omega t - k_x x)} \tag{1.11}$$

For medium 2:

$$E_2 = E_0 e^{-(k_x^2 - \varepsilon_2 \frac{\omega^2}{c^2})^{1/2} z} e^{i(\omega t - k_x x)} \tag{1.12}$$

with $(k^{2x} - \varepsilon_1 \frac{\omega^2}{c^2})^{1/2}$ and $(k_x^2 - \varepsilon_2 \frac{\omega^2}{c^2})^{1/2} > 0$ in order for the field to decay exponentially and being $\varepsilon_1 \equiv n_1^2$ and $\varepsilon_2 \equiv n_2^2$ the complex dielectric constants of the dielectric and metallic media, respectively.

The boundary conditions require continuity of the tangential components of electric field, $\vec{E}$, and magnetic field, $\vec{H}$, at $z = 0$. For that, the following constraint is necessary:

$$\left(k_x^2 - \varepsilon_1 \frac{\omega^2}{c^2}\right)^{1/2} = -\left(k_x^2 - \varepsilon_2 \frac{\omega^2}{c^2}\right)^{1/2} \tag{1.13}$$

This expression can only be true if $\varepsilon_2 < 0$ (i.e., a metal) and from Eq. 1.13, the dispersion relation for surface waves is obtained:

$$k_{SP} = k_x = \frac{\omega}{c}\sqrt{\frac{\varepsilon_1 \varepsilon_2}{\varepsilon_1 + \varepsilon_2}} \tag{1.14}$$

being valid for $|\varepsilon_2| < \varepsilon_1$, remembering that medium 1 is the dielectric medium and medium 2 is the metallic medium.

When dielectric medium 1 is air ($\varepsilon_1 \simeq 1$) and medium 2 is a metal, the dielectric function of free electron gas for metals is given by:

$$\varepsilon_2(\omega) = 1 - \left(\frac{\omega_p}{\omega}\right)^2 \tag{1.15}$$

where ω_p is the plasma frequency [21] (see Sect. 1.3).

Using Eqs. 1.14 and 1.15, the dispersion relationship for surface waves at the metal/air interface can be expressed by:

$$\omega^2 = (ck_x)^2 \left[1 + \frac{1}{1 - (\omega_p^2/\omega)^2}\right] = \frac{\omega_p^2}{2} + (ck_x)^2 - \sqrt{\frac{\omega_p^4}{4} + (ck_x)^4} \tag{1.16}$$

For small k_x ($ck_x \ll \omega_p$), Eq. 1.16 becomes:

$$\omega^2 \simeq (ck_x)^2 \left[1 - \frac{1}{2}\left(\frac{ck_x}{\omega_p}\right)^2\right] \rightarrow \omega_{SP} \simeq ck_x \tag{1.17}$$

and for large k_x ($ck_x \gg \omega_p$), Eq. 1.16 becomes:

$$\omega^2 \simeq \frac{\omega_p^2}{2}\left[1 - \frac{1}{4}\left(\frac{\omega_p}{ck_x}\right)^2\right] \rightarrow \omega_{SP} \simeq \sqrt{\frac{\omega_p}{1 + \varepsilon_1}} \tag{1.18}$$

For values $k_x > 0$, SPs propagate along the metal/air interface placing on the right of the dispersion relation for light in air, $\omega = ck_x$ (or other dielectric medium, $\omega = ck_x/\sqrt{\varepsilon_1}$) as Fig. 1.15 illustrates. When $\omega < \omega_p/\sqrt{1+\varepsilon_1}$, SPs cannot transform into light and so, SPs are non radiative waves that travel in the surface. For $\omega > \omega_p$, the wavevectors (k_x and k_z) take real values and the SPs radiate out of surface and they are called radiative SPs [35].

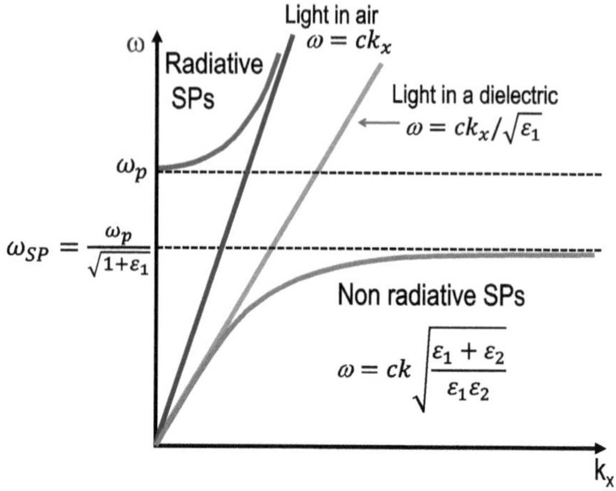

Fig. 1.15 Dispersion relation curves for air, any dielectric and SPs at interface between a dielectric medium (ε_1) and a metallic medium (ε_2). Note that the *SP curve* does not cross with that of light in air. Therefore, SPs cannot be directly excited by light propagating from air to a metallic medium [34]

1.3.2.3 Propagating Length and Penetration Depth

SPs propagate with high attenuation in the visible and the near infrared spectrum regions due to the high loss in the metal (Joule effect). Since the distribution of the electromagnetic field is not symmetric with respect to the interface, mostly the field is concentrated in the dielectric [5, 36]. SPs can propagate around a few of micrometers at visible wavelength and upto 100 μm in the near infrared [36–38]. The propagation length of SPs along the dielectric/metal interface is given by the following expression [5, 36]:

$$L_{SP} = \frac{1}{2Im(k_{SP})} \tag{1.19}$$

where $Im(k_{SP})$ represents the imaginary part of the wavevector of SPs.

The penetration depth or decay length explains the sensitivity at the interface of the evanescent field. Only changes of the dielectric properties in the vicinity of the interface between the dielectric and metal will affect the evanescent field and they will be observed [33]. The decay into the metal is much shorter than into the dielectric, as shown Fig. 1.16, which depends on the resonance wavelength. The penetration depth into the dielectric medium is of the order of half wavelength while into the metal is of a few tens of nanometers and is comparable to the skin depth [33, 38, 39]. The penetration depth can be calculated from [40, 41]:

For medium 1 (dielectric medium):

$$\delta_{SP1} = \delta_d = \frac{1}{k_{z1}} = \frac{c}{\omega}\sqrt{\frac{\varepsilon_1 + Re(\varepsilon_2)}{\varepsilon_1^2}} \tag{1.20}$$

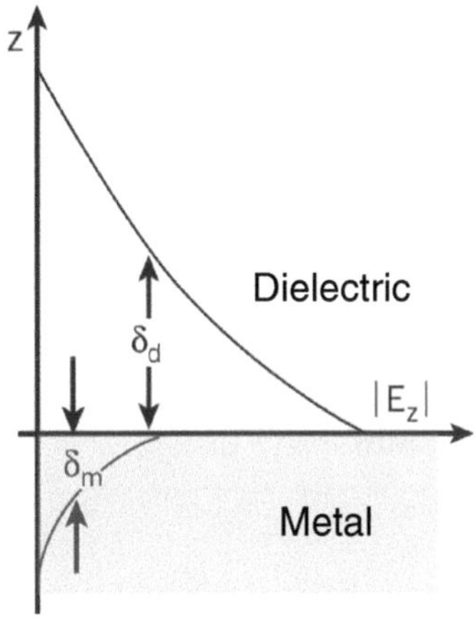

Fig. 1.16 Penetration depth of the evanescent field in the dielectric and metallic media. Figure adapted from Barnes et al. [42]

For medium 2 (metallic medium):

$$\delta_{SP2} = \delta_m = \frac{1}{k_{z2}} = \frac{c}{\omega}\sqrt{\frac{\varepsilon_1 + Re(\varepsilon_2)}{Re(\varepsilon_2)^2}} \tag{1.21}$$

where $Re(\varepsilon_2)$ represents the real part of the dielectric function of the metal.

1.3.2.4 Excitation of Surface Plasmons

SPs can be excited by electrons or light. However, the excitation using light is experimentally more affordable [5, 43]. As shown in Fig. 1.15, the dispersion relation of SPs and that of the incident electromagnetic wave (light) in air do not cross each other, thus SPs cannot be excited when the light is incident from air. Moreover, when it is necessary to illuminate the metal/air interface at a certain angle, the situation is even worse since the wavevector of the light in the plane has to be multiplied by $\sin\theta$, $k_x = k \sin\theta$, and it is, then, smaller than that of the incident electromagnetic waves parallel to the metal/air interface $k_x = k$ (see Fig. 1.17). Therefore, a direct excitation of SPs from air is not possible and several experimental techniques have been developed to provide the necessary wavevector conservation: prism coupling in attenuated total reflectance (ATR) conditions [15, 16, 42, 44], grating coupling [10, 14], waveguides using subwavelength protrusions or holes [45, 46], etc.

- **Prism coupling**

As explained above, SPs cannot be directly excited by incident light from air since the dispersion relations of SPs and that of evanescent field of the incident light do not cross. In order to match the frequency, ω, and the wavevector, k_x, of the SPs to those of the incident radiation, the wavevector of the incident light must be increased. For that, it is possible to place a dielectric medium with a refractive index $n > 1$, such as

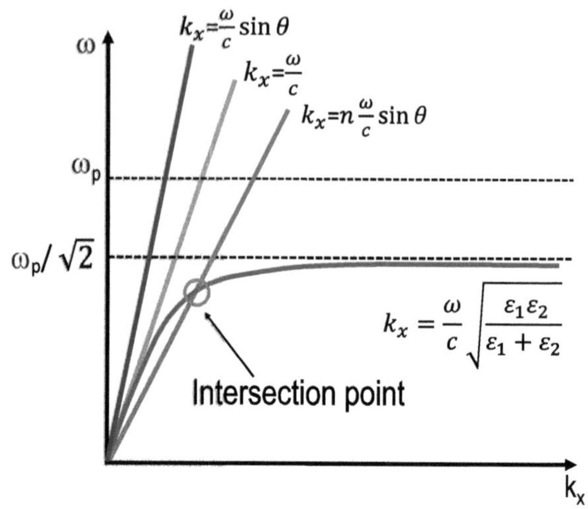

Fig. 1.17 Dispersion relation curves for SPs, light in air ($k_x = \frac{\omega}{c}$), light in air at a certain incident angle ($k_x = \frac{\omega}{c}sin\theta$) and light in a dielectric medium such as a glass at a certain incident angle ($k_x = \frac{\omega}{c}nsin\theta$). Note that the *SP curve* crosses with that of the light passing for a dielectric medium with a refractive index n (prism)

a glass. In this way, the wavevector, k_x, of the incident light will increase in a factor n and the dispersion relations of the SPs and the evanescent field of the incident light will intersect each other, as shown Fig. 1.17.

There are different configuration using prism coupling, as dielectric medium with a refractive index $n > 1$, to excite SPs with incident light: the Otto [15], the Kretschmann-Raether [16] and the Sarid [44] configurations, being the first two the most employed (see Fig. 1.18). These configurations use the ATR in a prism, as medium with a high refractive index, which is placed near or coupled to a metal surface. ATR is a technique that uses the total internal reflection[1] to produce evanescent waves (see Fig. 1.14). The use of a prism involves the coupling of the SPs and the evanescent field, which is formed upon total internal reflection of the incident light at the prism/metal interface. The evanescent field of the incident light travels through the metal (from the prism) up to the metal/dielectric (air) interface, where the SPs can be excited as shown Fig. 1.17 (intersection point), verifying the resonance condition:

$$k_{SPs} = k_x = \frac{\omega}{c}nsin\theta = \frac{\omega}{c}\sqrt{\frac{\varepsilon_1\varepsilon_2}{\varepsilon_1+\varepsilon_2}} \tag{1.22}$$

From the Eq. 1.22, it is possible to calculate the incident angle corresponding to the coupling between the evanescent field of the incident light and the SPs:

$$\theta_{SPs} = arcsin\sqrt{\frac{\varepsilon_1\varepsilon_2}{(\varepsilon_1+\varepsilon_2)\varepsilon_p}} \tag{1.23}$$

where the dielectric constant of prism is given by $\varepsilon_p = n^2$.

[1]The total internal reflection occurs when the incident angle of the light is higher than the critical angle, $\theta_c = \sin^{-1}\frac{n_2}{n_1}$, which is the minimum angle for the total internal reflection occurs. In this way, the light is totally reflected [47], penetrating into the medium with the higher refractive index.

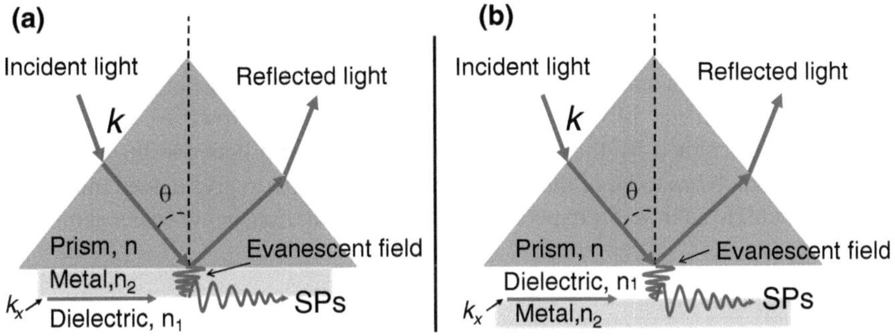

Fig. 1.18 **a** Kretschmann-Raether and **b** Otto configurations for the excitation of SPs at a metal/dielectric interface, where $k_x = knsin\theta$ being n the refractive index of the prism, k the light wavevector and θ the incident angle of the light

∘ *Kretschmann – Raether configuration*

In the Kretschmann-Raether configuration [16], a metal film is coupled to a glass prism. In this case, the film is illuminated through the prism at an incident angle $\theta > \theta_c$ and the wavevector of the incident light is increased. The evanescent field of the incident light, created at the prism/metal interface, propagates through the metal film reaching the metal/dielectric interface. At a certain incident angle, the in-plane wavevector of the evanescent field crosses with that of the SPs at the metal/dielectric interface and the SPs are excited [33], as shown Fig. 1.18a.

Under the resonance conditions, a minimum is observed in the reflected light when the evanescent field of the incident light is coupled to SPs. However, this method is very limited by the thickness of metal film to a narrow window $\sim 50 \pm 10$ nm, if the optimal absorbance is required [5, 48]. For films thicker than 80 nm, the evanescent field of the incident light reaching the other side of the metallic film is very weak because it decays exponentially with the distance. For thinner films than 40 nm, the electron oscillations are damped due to confinement effects [5, 48]. Despite this, SPs can be excited easily and this configuration is widely accepted. Thus, the Kretschmann-Raether configuration is employed in this work and is explained in detail in the following chapter (see Chap. 2).

∘ *Otto configuration*

In the Otto configuration [15], a metal surface is placed near a glass prism, where the metal surface and the prism are separated a certain distance through a dielectric medium, as shown Fig. 1.18b. In this case, the prism/dielectric interface is illuminated with light at a certain angle $\theta > \theta_c$ and the evanescent field of the incident light propagates through the dielectric medium reaching the dielectric/metal interface. At a certain incident angle, the in-plane wavevector of the field evanescent crosses with that of the SPs for that frequency at the dielectric/metal interface and the SPs are excited [33], as Fig. 1.18b illustrates.

Within this method, a minimum is also observed in the reflected light when the evanescent field of the incident light is coupled to SPs. The Otto configuration is suitable to excite plasmons in metal crystal surfaces and it is not limited by the thickness of the metal film as in the prior configuration. However, the coupling strength is determined by the width of dielectric medium between the prism and metal film, which has to be of the order of half wavelength to give a deep minimum. This is difficult to achieve for experiments at visible radiation, being easier at infrared radiation.

∘ *Sarid configuration*

The Sarid configuration [44] is similar the Otto configuration [15] but using a metal thin film of the order of 20 nm [49]. In this case, it is possible to excite a long-range SP (symmetric mode) and a short-range SP (asymmetric mode), while in the Kretschmann-Raether [16] and the Otto [15] configuration just short-range SPs can be excited.

- **Grating coupling**

SPs can be excited using diffraction effects at a grating pattern on a metal surface, as Fig. 1.19 shows. A rough metal surface corresponds to a superposition of different gratings, thus SPs can propagate at the rough interface between the metal and a dielectric medium with a refractive index n [50].

Incident light at an incident angle θ with respect to the normal of the average plane of the rough surface can be diffracted, increasing or decreasing the wavevector $k_x = k \sin\theta$ and forming a series of diffracted modes. SPs can be excited if the incident light wavevector along the interface of a diffracted order is equal to that of SPs, according to the coupling condition [50]:

$$k_{SPs} = nk\sin\theta + k_{grating} = nk\sin\theta \pm mG \tag{1.24}$$

where $m = 1, 2, 3\ldots$ and $G = \frac{2\pi}{b}$ is the reciprocal vector of the grating, being b the lattice constant.

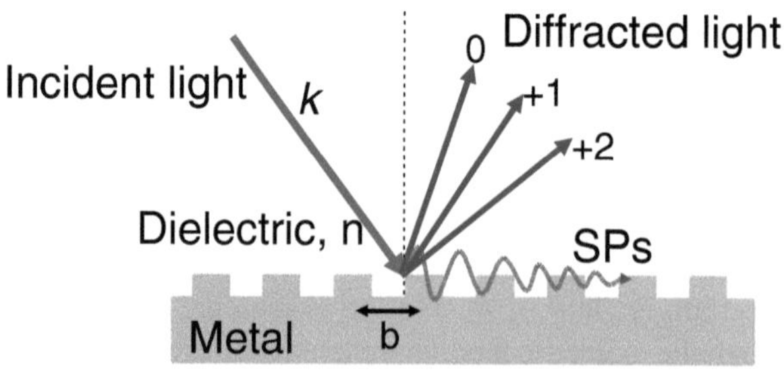

Fig. 1.19 Grating coupling of the incident light with the wavevector k_x on a metal grating surface of period b

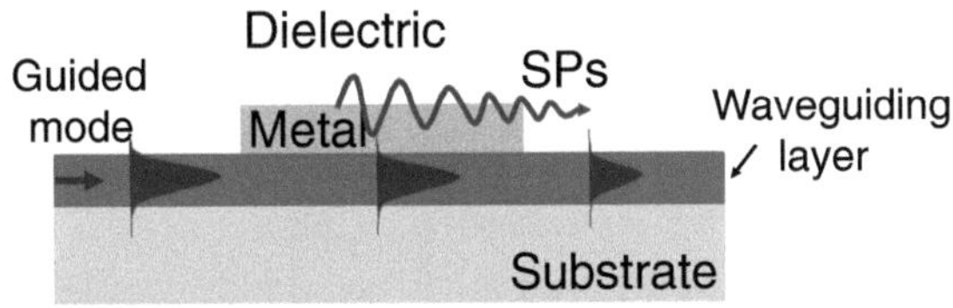

Fig. 1.20 Scheme of a waveguide coupling of SPs with light waveguide modes at the metal/dielectric interface

By this configuration, the excitation of non radiative SPs is also detected as a minimum in the reflected light.

- **Waveguide coupling**

Waveguides can manipulate and propagate light at the nanoscale, and can be used in addition to excite SPs by modes in a dielectric waveguide [42, 43].

In this case, light guided modes propagate along a waveguide using the total internal reflection, generating an evanescent field at the waveguide/metal interface. This evanescent field propagates through the metal film and can be coupled to SPs [51], as Fig. 1.20 shows. SPs are excited when the wavevector of the evanescent field intersects with that of the SPs at the metal/dielectric interface. Optical fibers can be used to excite the SPs in an easy and flexible way and can be easily integrated with other optical and electrical components [42, 43, 51].

1.3.2.5 Field Enhancement

The electromagnetic field of the light is much more intense at the metal/dielectric interface where the reflected light shows its lowest value. If the dielectric medium is air, $\varepsilon_1 \simeq 1$, the enhancement value is given by the ratio of the field intensity on the metal surface at the air side ($|H_z(1/2)|^2$) divided by the incoming field intensity ($|H_{z0}(1/2)|^2$) [5]:

$$\frac{|H_z(1/2)|^2}{|H_{z0}(1/2)|^2} = \frac{t_{02}^p t_{21}^p e^{ik_{z2}d}}{1 + r_{02} r_{21} e^{2ik_{z2}d}} \tag{1.25}$$

where t_{ik}^p correspond to the Fresnel transmission coefficients being the medium 0 = prism, medium 2 = metal film and medium 1 = air, k_{2z} is the wavevector of evanescent field perpendicular to the surface and d is the metal film thickness.

This behavior can be observed in Fig. 1.21a, which presents the electromagnetic field as a function of the distance from the prism in z direction at three different incident angles. At resonant conditions, 45.2°, the electromagnetic field associated with SPs is much larger than the evanescent field of the incident light, so the final field intensity decreases from the air towards the prism. Outside the resonant conditions, 50°, the only appreciable electromagnetic field inside the metal is the evanescent field of the incident light, which decreases exponentially from the prism towards the air. Finally, for an intermediate situation, 45.4°, both fields may have similar values and the intensity profile shows a minimum inside the metal layer [5].

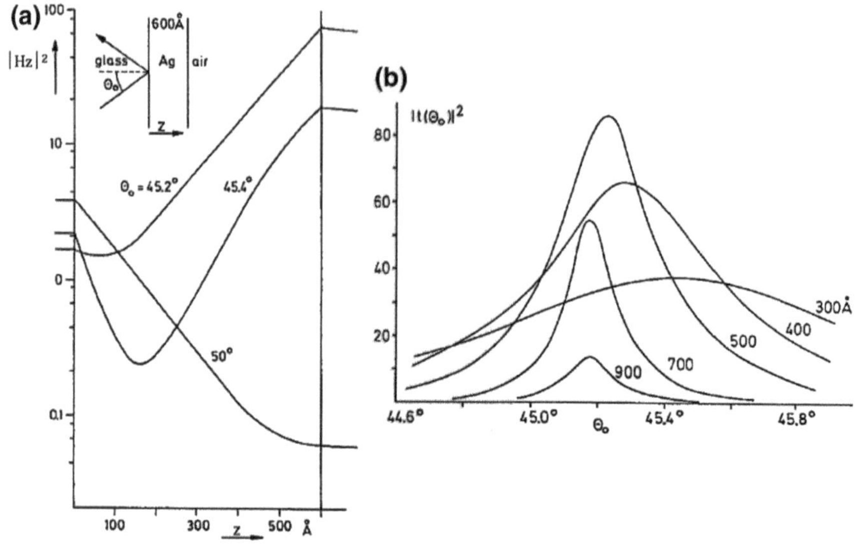

Fig. 1.21 **a** Intensity of the relative electromagnetic field as a function of the distance from the prism towards the air and **b** enhancement factor of Ag thin film with different thickness indicated in the figure. Figure adapted from Raether et al. [5]

Moreover, the field enhancement also depends strongly on the incident angle. Figure 1.21b shows that at resonance conditions, the enhancement factor of the field can reach values up to 80 for Ag. Au films take values up to 50 [48].

1.4 Surface Plasmons Today

SPs are employed in a wide range of fields such as energy [52–54], biomedicine [4, 27, 55, 56], environment [57, 58], information technology [3, 42, 59], nanoelectronic [60] and sensors [61, 62].

Nowadays, the growth of plasmonics is clearly reflected in the literature, as shown Fig. 1.2. Despite SPs have been known for over a century, the number of related publications has rapidly grown in the last years due to the development of powerful tools for fabricating nanomaterials in a controlled way and measuring SPR, with the ability to manipulate the light at the subwavelength scale [2, 62].

SPs show interesting features and their applications are mainly based on two properties. On the one hand, as described above, SPs allow overcoming the diffraction limit of traditional optics and can concentrate and amplify the light electric field at subwavelength scale in more than one order of magnitude, increasing the resolution and the sensitivity of optical probes and optoelectronic devices [42, 63, 64]. On the other hand, SP is as well a very weakly attenuated resonance, so SPs are highly sensitive to modifications in the surrounding media close to the metallic nanostructures. Thus, usually, SPR devices are used as sensors in a wide range of fields [4, 61, 62]. The first sensor based on SPR was reported in 1983 [65]. Since then, sensors based on SPR have been developed and used to identify elements or for analyzing some molecular interactions. For example, BIAcore technology uses SPR to investigate specific interactions at the surface of a sensor chip [66]. Furthermore as sensor, SPR is widely used as probe to study the features of metallic nanostructures in many different approaches as well as the characteristics of the adjacent dielectric media [67–69].

Nowadays, research and development have been carried out to incorporate the SPR technology into many different fields, in order to use the unique properties of the SPs [33]. Thanks to the development of the SPR technology, this phenomenon may be incorporated to other techniques and may be used to analyze diverse systems. However, despite rapid growth of the field of plasmonics in the last years, large amount of work remains to be done.

1.5 Aims of This Work

The extreme sensitivity of SPR to the features of metal nanostructures and surrounding dielectric media, makes SPs are an extraordinarily useful tool to characterize materials in different fields [67–69]. However, the use of SPs to follow the evolution of modifications in nanostructures has not been almost explored. Usually, this spectroscopy is limited to study characteristics of materials or to analyze molecular interactions. Thus, in this thesis we present a set of experiments where SPs are used to follow modifications induced in nanomaterials based on Au under different ways, namely:

- Structural modifications of Au films and Au/Fe bilayers annealed under different conditions for promoting their transition from a continuous film to nanoparticles, modulating their optical properties. In the case of Au films, we can follow the transition from ESPs to LSPs.
- Modifications induced on matter using hard X-ray irradiation, specifically, on glasses and organic films (Co-phthalocyanine films) in contact with a Au film. For this purpose, we designed, mounted and tested a new setup that allows measurements of SPR in situ and in real time, while the samples are being irradiated with hard X-rays in a synchrotron beamline (SPR-XAS setup).

The particular interest and potential applications of each proposed experiment and analyzed system are addressed in the corresponding chapter of this thesis work.

In order to achieve the aims exposed here, a careful and systematic work is carried out both for the fabrication of samples and for their characterization.

1.6 Organization of the Work

This work is divided in eight chapters. This chapter shows a review of the surface plasmons both in nanoparticles and metallic films, the current use of SPs and the objectives of this work. In the following chapter (Chap. 2), the fabrication of samples and the experimental techniques employed for their characterization are detailed. Specifically, great attention is placed on the experimental setup based on the Kretschmann-Raether configuration to measure SPR. In Chap. 3, variations of SPs in Au thin films annealed in air at different temperatures are studied, changing the initial film thickness. In Chap. 4, we show the fabrication of Au/FeO_x NPs over silica substrates from thin films and bilayers evaporated and treated thermally. Subsequently, SPR modifications are studied changing the film thickness, configuration of samples and annealing conditions. Chapter 5 presents the design and development of a SPR system based on the Kretschmann-Raether configuration, compatible with an X-ray absorption spectroscopy beamline (SPR-XAS setup). In Chap. 6, a study of soda-lime and silica glasses irradiated with X-rays by the SPR-XAS setup is presented, analyzing the possible modifications on the optical properties of the glass substrates. In Chap. 7, effects of X-ray irradiation on the optical properties of CoPc layers, grown at RT and 200 °C onto Au film/silica substrate system, are evaluated using the SPR-XAS setup. In all cases, a deep characterization of samples is carried out using different microscopies and spectroscopies. Finally, in Chap. 8, the most significant conclusions of this work are summarized.

References

1. M.V. Uwe Kreibig, *Optical Properties of Metal Clusters, Springer Series in Material Science* (Springer, Berlin, 1995)
2. M.L. Brongersma, P.G. Kik, *Surface Plasmon Nanophotonics* (Spring, Berlin, 1988)
3. S.A. Maier, *Plasmonics* (Springer, Berlin, 2006)
4. M.A. Garcia, in *Recent Developments*, ed. by A. Tiwari, S. Pilla. Bio-Nanocomposites for Biomedical Applications, Surface Plamons in biomedicine, 2010
5. H. Raether, *Surface Plasmons on Smooth and Rouge Surfaces and on Gratings* (Springer, Berlin, 1988)
6. J.C.M. Garnett, Colours in metal glasses and in metallic films. Philos. Trans. R. Soc. Lond. Ser. A, Containing Pap. Math. Phys. Char. **203**(359–371), 385–420 (1904)
7. G. Mie, Beiträge zur Optik trüber Medien, speziell kolloidaler Metallösungen. Annalen der Physik **330**(3), 377–445 (1908)
8. S. Eustis, M.A. El-Sayed, Determination of the aspect ratio statistical distribution of gold nanorods in solution from a theoretical fit of the observed inhomogeneously broadened longitudinal plasmon resonance absorption spectrum. J. Appl. Phys. **100**(4), 044324 (2006)
9. H. DeVoe, Optical properties of molecular aggregates. i. classical model of electronic absorption and refraction. J. Chem. Phis. **41**, 393–400 (1964)
10. R.W. Wood, On a remarkable case of uneven distribution of light in a diffraction grating spectrum. Proc. Phys. Soc. London **18**(1), 269 (1902)
11. U. Fano, The theory of anomalous diffraction gratings and of quasi-stationary waves on metallic surfaces (Sommerfeld's Waves). J. Optical Soc. Am. **31**(3), 213–222 (1941)

12. D. Pines, Collective energy losses in solids. Rev. Mod. Phys. **28**, 184–198 (1956)
13. C.J. Powell, J.B. Swan, Origin of the characteristic electron energy losses in magnesium. Phys. Rev. **116**, 81–93 (1959)
14. R.H. Ritchie, E.T. Arakawa, J.J. Cowan, R.N. Hamm, Surface-plasmon resonance effect in grating diffraction. Phys. Rev. Lett. **21**, 1530–1533 (1968)
15. A. Otto, Excitation of nonradiative surface plasma waves in silver by the method of frustrated total reflection. Zeitschrift für Physik **216**, 398–410 (1968)
16. H. Raether, E. Kretschmann, Radiative decay of non radiative surface plasmons excited by light. Zeitschrift für Naturforschung A **23**, 2135–2136 (1968)
17. E. Kretschmann, Die bestimmung optischer konstanten von metallen durch anregung von oberflächenplasmaschwingungen. Zeitschrift für Physik **241**(4), 313–324 (1971)
18. U. Kreibig, P. Zacharias, Surface plasma resonances in small spherical silver and gold particles. Zeitschrift für Physik **231**, 128–143 (1970)
19. S.L. Cunningham, A.A. Maradudin, R.F. Wallis, Effect of a charge layer on the surface-plasmon-polariton dispersion curve. Phys. Rev. B **10**, 3342–3355 (1974)
20. M. Fleischmann, P.J. Hendra, A.J. Mcquillan, Raman spectra of pyridine adsorbed at a silver electrode. Chem. Phys. Lett. **26**(2), 163–166 (1974)
21. C. Kittle, *Introduction to Solid State Physics*, 8th edn. (Wiley, New York, 2005)
22. C.D. Geddes (ed.), *Reviews in Plasmonics 2010*. (Springer, New York, 2012)
23. M.A. Garcia, Surface plasmons in metallic nanoparticles: fundamentals and applications. J. Phys. D: Appl. Phys. **44**(28), 283001 (2011)
24. N.W. Ashcroft, N.D. Mermin, Solid State Phys. (1976)
25. J. de la Venta Granda, Propiedades magnéticas de nanoparticulas de oro. PhD thesis, Universidad Complutense de Madrid, Febrero, 2009
26. C. Noguez, Surface plasmons on metal nanoparticles: the influence of shape and physical environment. J. Phys. Chem. C **111**, 3806–3819 (2007)
27. M. Hu, J. Chen, Z.Y. Li, L. Au, G.V. Hartland, X. Li, M. Marquez, Y. Xia, Gold nanostructures: engineering their plasmonic properties for biomedical applications. Chem. Soc. Rev. **35**(11), 1084–1094 (2006)
28. E. Hutter, J.H. Fendler, Exploitation of localized surface plasmon resonance. Adv. Mater. **16**(19), 1685–1706 (2004)
29. P. Zhang, T.K. Sham, Fabrication of thiol-capped pd nanoparticles: an electrochemical method. Appl. Phys. Lett. **82**(11), 1778–1780 (2003)
30. M.A. García, J. Llopis, S.E. Paje, A simple model for evaluating the optical absorption spectrum from small au-colloids in sol-gel films. Chem. Phys. Lett. **315**(5–6), 313–320 (1999)
31. M. Quinten, U. Kreibig, Absorption and elastic scattering of light by particle aggregates. Appl. Opt. **32**(30), 6173–6182 (1993)
32. F.J. Garcia-Vidal, L. Martín-Moreno, J.B. Pendry, Surfaces with holes in them: new plasmonic metamaterials. J. Opt. A: Pure Appl. Opt. **7**(2), S97 (2005)
33. R.B.M. Schasfoort, A.J. Tudos (eds.), *Handbook of Surface Plasmon Resonance*. (The Royal Society of Chemistry, Cambridge, 2008)
34. J. Stoltenberg, D. Pengra, Surface Plasmon Resonance in a Thin Metal Film, Master's thesis, 2008
35. H. Raether, Significative of surface plasmons for surface studies. Journal de Physique Colloques **31**, 59–62 (1970)
36. M. Kuttge, Cathodoluminescence plasmon microscopy. PhD thesis, Universiteit Utrecht, 2009
37. M. Kuttge, E.J.R. Vesseur, J. Verhoeven, H.J. Lezec, H.A. Atwater, A. Polman, Loss mechanisms of surface plasmon polaritons on gold probed by cathodoluminescence imaging spectroscopy. Appl. Phys. Lett. **93**(11), 113110 (2008)
38. P.A. de Carvalho Gomes, Surface Plasmon Resonance as a Tool in the Functional Analysis of an Immunodominant Site in Foot-and-Mouth Disease Virus. PhD thesis, Universitat de Barcelona, 2003
39. Z. Han, S.I. Bozhevolnyi, Radiation guiding with surface plasmon polaritons. Rep. Prog. Phys. **76**(1), 016402 (2013)

40. R.H. Ritchie, Plasma losses by fast electrons in thin films. Phys. Rev. **106**, 874–881 (1957)
41. X. Lu, Studies of surface plasomns and localized surface plasmons at teraherzs frequencies. PhD thesis, Tianjin University, 2003
42. W.L. Barnes, A. Dereux, T.W. Ebbesen, Surface plasmon subwavelength optics. Nature **424**(6950), 824–830 (2003)
43. S. Zuccon, Plasmonics and near field optics for sensing purposes. PhD thesis, University of Padova, 2010
44. D. Sarid, Long-range surface-plasma waves on very thin metal films. Phys. Rev. Lett. **47**(26), 1927–1930 (1981)
45. B. Hecht, H. Bielefeldt, L. Novotny, Y. Inouye, D.W. Pohl, Local excitation, scattering, and interference of surface plasmons. Phys. Rev. Lett. **77**, 1889–1892 (1996)
46. S.A. Maier, P.E. Barclay, T.J. Johnson, M.D. Friedman, O. Painter, Low-loss fiber accessible plasmon waveguide for planar energy guiding and sensing. Appl. Phys. Lett. **84**(20), 3990–3992 (2004)
47. S.S. Bokatzian-Johnson, Surface Plasmon Resonance: Principles and Applications, Master's thesis, University of Alabama, 2008
48. W. Knoll, Interfaces and thin films as seen by bound electromagnetic waves. Ann Rev. Phys. chem. **49**, 569–638 (1998)
49. O. Mata-Mendez, J. Fonseca-Camarillo, Excitaci6n de plasmones de superficie de largo alcance con la configuraci6n de sarid: influencia del haz incidente. Revista Mexicana de Física **43**(3), 451–467 (1997)
50. A.P. Hibbins, Grating Copuplong of Surface Plasmon Polaritons at Visible and Microwave Frequencies. PhD thesis, University of Exeter, 1999
51. X. Fan, I.M. White, S.I. Shopova, H. Zhu, J.D. Suter, Y. Sun, Sensitive optical biosensors for unlabeled targets: a review. Anal. Chim. Acta **620**(1–2), 8–26 (2008)
52. S. Pillai, K.R. Catchpole, T. Trupke, M.A. Green, Surface plasmon enhanced silicon solar cells. J. Appl. Phys. **101**(9), 093105 (2007)
53. H. Atwater, A. Polman, Plasmonics for improved photovoltaic devices. Nat. mater. **9**(3), 205–213 (2010)
54. P. Matheu, S.H. Lim, D. Derkacs, C. McPheeters, E.T. Yu, Metal and dielectric nanoparticle scattering for improved optical absorption in photovoltaic devices. Appl. Phys. Lett. **93**(11), 113108 (2008)
55. W. Fritzsche, T.A. Taton, Metal nanoparticles as labels for heterogeneous, chip-based dna detection. Nanotechnology **14**(12), R63 (2003)
56. S. Eustis, M. El-Sayed, Why gold nanoparticles are more precious than pretty gold: noble metal surface plasmon resonance and its enhancement of the radiative and nonradiative properties of nanocrystals of different shapes. Chem. Soc. Rev. **35**(3), 209–217 (2006)
57. R. Narayanan, M.A. El-Sayed, Changing catalytic activity during colloidal platinum nanocatalysis due to shape changes: electron-transfer reaction. J. Am. Chem. Soc. **126**(23), 7194–7195 (2004)
58. K. Awazu, M. Fujimaki, C. Rockstuhl, J. Tominaga, H. Murakami, Y. Ohki, N. Yoshida, T. Watanabe, A plasmonic photocatalyst consisting of silver nanoparticles embedded in titanium dioxide. J. Am. Chem. Soc. **130**(5), 1676–1680 (2008)
59. E. Ozbay, Plasmonics: merging photonics and electronics at nanoscale dimensions. Science **311**(5758), 189–193 (2006)
60. M. Dragoman, D. Dragoman, Nanoelectronics Principles and Devices, ch. 1, pp. 13–18. 2006
61. J. Homola, S.S. Yee, G. Gauglitz, Surface plasmon resonance sensors: review. Sens. Actuators B: Chem. **54**, 3–15 (1999)
62. B.D. Gupta, R.K. Verma, Surface plasmon resonance-based fiber optic sensors: principle, probe designs, and some applications. J. Sens. **2009**, 1–12 (2009)
63. M. Dragoman, D. Dragoman, Plasmonics: applications to nanoscale terahertz and optical devices. Prog. Quantum Electron. **32**, 1–41 (2008)
64. M. Heo, H. Cho, J. Jung, J.R. Jeong, S. Park, J.Y. Kim, High-performance organic optoelectronic devices enhanced by surface plasmon resonance. Adv. Mater. **23**(47), 5689–5693 (2011)

65. B. Liedberg, C. Nylander, I. Lunström, Surface plasmon resonance for gas detection and biosensing. Sens. Actuators **4**, 299–304 (1983)
66. See https://www.biacore.com/lifesciences/index.html
67. J. Homola, Surface plasmon resonance sensors for detection of chemical and biological species. Chem. Rev. **108**(2), 462–493 (2008)
68. R.P.H. Kooyman, L.M. Lechuga, In *Immunosensors Based on Total Internal Reflectance*. Handbook of Biosensors and Electronic Noses: Medicine, Food and the Environment (1997)
69. A. Serrano, C. Monton, I. Valmiansk, F. Gálvez, A.L. Cortajarena, J. de la Venta, I.K. Shuller, M.A. García, Study of Co-phthalocyanine films by surface plasmon resonance spectroscopy. J. Appl. Phys. **115**, 103106 (2014)

Chapter 2
Experimental Techniques

The fabrication of samples and the experimental techniques employed for their characterization are detailed in this chapter. Particular attention is paid to the experimental setup based on the Kretschmann-Raether configuration for measuring surface plasmons in thin films.

2.1 Introduction

This chapter describes in detail the preparation of samples and experimental techniques used to characterize them. Initially, the methods used for the fabrication of samples are explained. Their final properties depend on the growth parameters and conditions. Later, both structural and spectroscopic experimental techniques are described. A good understanding of these techniques allows a correct analysis of the samples and the results. Particularly, greater attention is paid to the designed and mounted system to measure surface plasmon resonance in metallic thin films under the Kretschmann-Raether configuration [1].

2.2 Sample Growth

In this work, thin films and bilayers were grown by thermal evaporation and some of the samples were treated thermally, in order to obtain different kinds of nanostructures. Next, the fabrication of the different samples is described in detail.

A. Serrano Rubio, *Modified Au-Based Nanomaterials Studied by Surface Plasmon Resonance Spectroscopy*, Springer Theses, DOI 10.1007/978-3-319-19402-8_2

2.2.1 Thin Film Growth by Thermal Evaporation

The thermal evaporation is a technique of physical vapor deposition (PVD) [2], which consists of heating a material at high temperature to increase its vapor pressure. The material vapor finally condenses on the substrate surface forming a thin film [3, 4].

Usually, in this technique, low pressures below 10^{-4} torr are used to avoid reaction between the vapor material and chamber atmosphere. At these low pressures, the mean free path of vapor atoms is the of same order as the vacuum chamber dimensions, so these particles travel in straight lines from the evaporation source towards the substrate [3]. This can originate shadowing phenomena with 3D objects, especially in those regions not directly accessible from the evaporation source. Besides, in thermal evaporation techniques, the average kinetic energy of vapor atoms reaching the substrate surface is low (i.e., for Au, where the $T_{evap} = 1400\,°C$, the average kinetic energy per atom is around 0.2 eV [3]) and thus, it is smaller than the binding energy of atoms in a material (of the order of 5–10 eV). It implies that the atoms reach the substrate surface with low energy, affecting the morphology of the films, and often obtaining low density films and with low adhesion on the substrate surface. When the binding energy is smaller than the average kinetic energy of vapor atoms, the formation of the thin films is more homogeneous [2, 3, 5].

The film growth is a complex process, where a number of steps occur at microscopic level: arrival of atoms onto the surface, adsorption and subsequent surface diffusion, nucleation, formation of new layers, etc. Each of these steps, shown in Fig. 2.1, depends on both the previous step and the parameters of deposition process: deposition rate, temperature, evaporated material, etc. All these steps determinate the growth and the morphology of evaporated films. Besides this, intrinsic parameters such as the interaction between the evaporated material and the substrate (i.e., interface energy) and the substrate cleaning, play an important role in the film morphology.

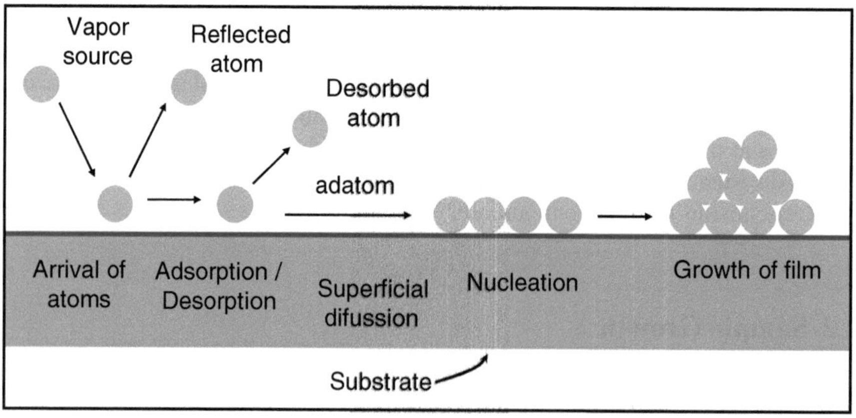

Fig. 2.1 Steps involved in the growth of thin films by PVD techniques

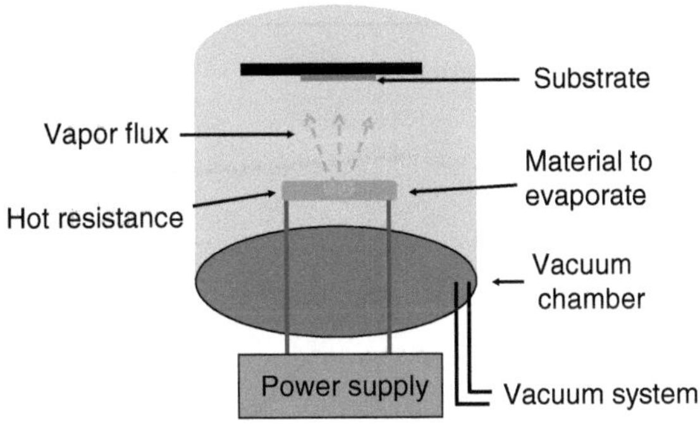

Fig. 2.2 Scheme of thermal evaporation by resistance heating

In thermal evaporation techniques, different methods can be applied to heat the source: by resistance heating or bombardment with a high energy electron beam (usually to several keV) from an electron beam gun [3]. In this work, all films were thermally evaporated by resistance heating, using a power supply of low voltage and high current to avoid the formation of electrical discharges between the contacts. Resistance heating is the process by which an electric current passes through a conductor, releasing heat (Joule effect, $Q \propto I^2 \cdot R$ being Q the amount of heat released, I the intensity and R the resistance). A simple scheme of thermal evaporation by resistance heating is shown in Fig. 2.2.

Different heating elements can be employed to evaporate by resistance heating. The most used is a spiral-shaped or "V"-shaped conductor filament (i.e., W, Mo), in order to generate deformed regions with a high concentration of defects, where the resistance increases (the heat increases) when the electric current passes through the filament. Usually the filament is used to evaporate material forming a wire and the electric current is increased, rising the temperature, until the wire is melted. For materials forming powders, metal boats or ceramic crucible surrounded by a filament are usually used. The powders are deposited in the boat or crucible and the electric current passes through the metal boat or filament, heating until the powder is evaporated. For materials that may react with the resistance or with a melting point close to that of the resistance, ceramic crucibles are used [3]. A scheme of different elements for evaporating by resistance heating is shown in Fig. 2.3. In this work,

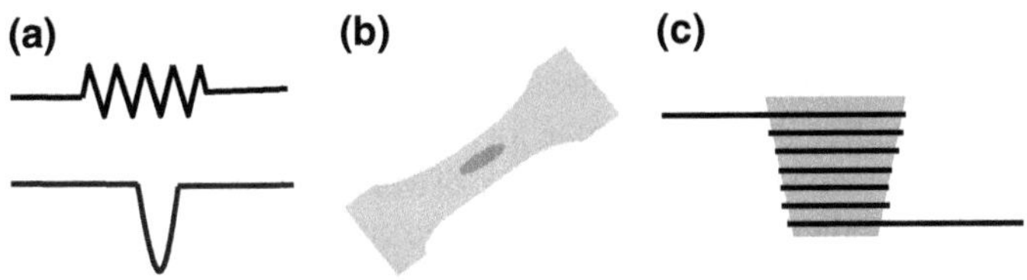

Fig. 2.3 Scheme of different heating elements to evaporate by resistance heating: **a** tungsten filaments, **b** metal boat and **c** ceramic crucible

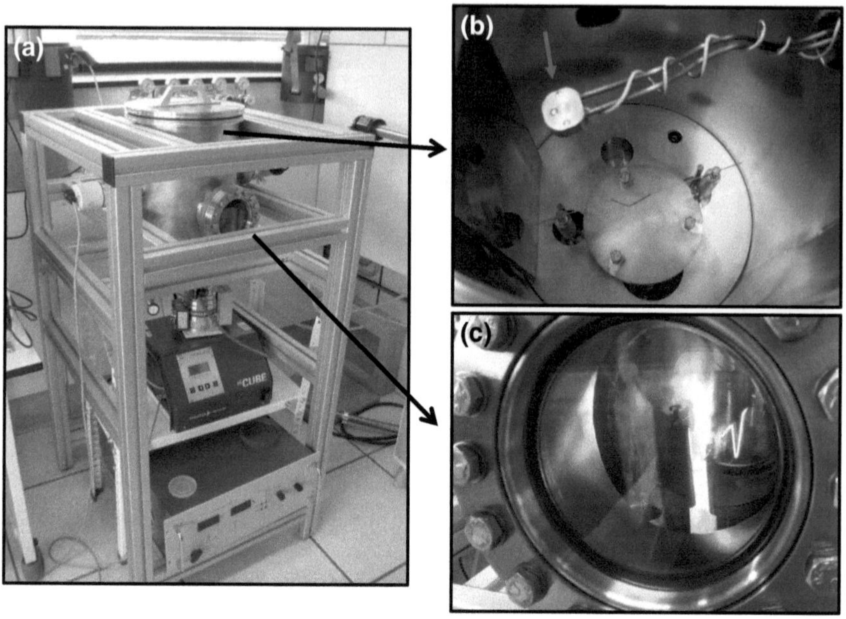

Fig. 2.4 Photographs of the home-made evaporation chamber mounted in the Departamento de Electrocerámica at the Instituto de Cerámica y Vidrio (CSIC) of Madrid: **a** front view, **b** view from top observing the chamber opened and **c** view from a glass window observing the deposition of a film

"V"-shaped tungsten filaments were used to grow metallic films from metal wires and a ceramic crucible was used to evaporate organic molecules.

In this work, samples were fabricated in different vacuum systems:

- A home-made evaporation chamber, which was designed and mounted in the Departamento de Electrocerámica at the Instituto de Cerámica y Vidrio (CSIC) of Madrid. Figure 2.4 displays some pictures of the vacuum system. This chamber contains, among other elements, of a large area where placing substrates of up to $20 \times 20\,\text{cm}^2$, a Q-microbalance (see Fig. 2.4b) in order to control the film thickness and a glass window to follow the deposition of the films on substrates (see Fig. 2.4c). The Q-microbalance is located close to the substrates in order to obtain thin films with similar thicknesses to those values detected by the microbalance, which is calibrated using X-ray reflectivity (see Sect. 2.6).
- Commercial Pfeiffer 306 auto coater located in the Departamento de Física de Materiales at the Universidad Complutense of Madrid. This chamber (see Fig. 2.5a) contains, among other elements, of a shutter to cover the substrates and to control parameters such as the time deposition or the homogeneity of the material vapor to deposit. The shutter is always open after a certain time, from which the possible impurities have been evaporated. Besides this, the system has a glass bell jar, which is very useful to follow the evaporation and see the deposition of the films

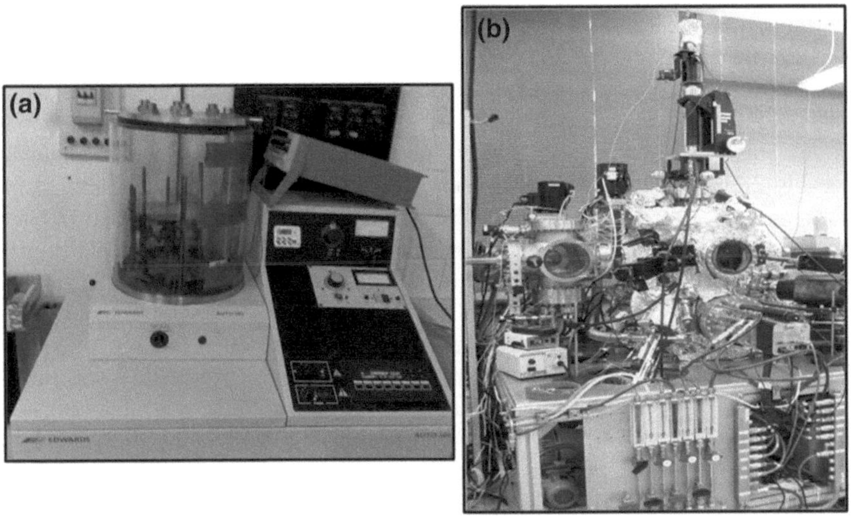

Fig. 2.5 Photographs of **a** commercial Pfeiffer 306 auto coater in the Departamento de Física de Materiales at the Universidad Complutense of Madrid and **b** the home-made evaporation chamber in the Schuller Nanoscience Group, at the University of California, San Diego

on substrates. With this chamber and the previous one, it is possible to grow metallic thin films (i.e., Au, Fe, Ag, Al films).

- A home-made evaporation chamber in the Schuller Nanoscience Group, at the University of California, San Diego. Figure 2.5b shows a picture of the vacuum system. This chamber consists of a Molecular Beam Epitaxy system and includes, among other elements, a Q-microbalance in order to control the thickness of the samples and a shutter to cover the substrates and to control the film deposition. Besides, with this chamber a homogeneous deposition can be achieved by a moderate rotation of substrate plate during the evaporation. With this chamber, it is possible to grow organic molecules in addition to metallic films.

Soda-lime and silica substrates were used in this work. Their features are the following:

- **Soda-lime**: commercial soda-lime glass substrates (Sigma-Aldrich) composed of SiO_2 73 %, Na_2O 14 %, CaO 7 %, MgO 4 %, Al_2O_3 2 %, with a thickness of 1 mm and a refractive index n = 1.513 ($\lambda = 645$ nm) [6].
- **Silica**: commercial silica glass substrates (UQG Optics) composed of SiO_2 99.9 % and Al_2O_3, Fe_2O_3, Na_2O, K_2O ppm, with a thickness of 1 mm and a refractive index n = 1.457 ($\lambda = 633$ nm) [7].

Prior to deposition of films, glass substrates were cleaned by a subsequent immersion in trichloroethylene, acetone, ethanol and distilled water and finally driedwith N_2 flux or using soap and water and then dried with dry air flux. Later, substrates were placed at a controlled distance from the evaporation source (see Fig. 2.2). Subsequently, Au films, Fe films, Fe/Au bilayers and Co-phthalocyanine (CoPc)/Au bilayers were thermally evaporated on the substrates, using resistance heating [3]. For thermal evaporation of Au and Fe films, Au or Fe wire (from Goodfellow) was placed onto a "*V*"-shaped tungsten filament in the vacuum chamber (see Fig. 2.2). The tungsten was chosen because presents a higher melting point ($T_{evap} = 3410\,°C$) than that of Au and Fe. The Au and Fe wires were initially melted to form a ball at the filament. Au films were grown using Au[1] wire of 99.99 % purity and a 0.5 mm diameter. Au films were fabricated with the Commercial Pfeiffer 306 auto coater in the Departamento de Física de Materiales at the Universidad Complutense of Madrid. The thickness of the Au films was controlled with the deposition time and a shutter to cover the samples of the Au evaporation source. This was calculated using the optical absorption of films and integrating in the range between 400 and 700 nm (see Sect. 2.6). Fe films were grown using Fe[2] wire of 99.99 % purity and a 0.5 mm diameter. Fe films and Fe/Au bilayers were grown using the homemade evaporation chamber in the Departamento de Electrocerámica at the Instituto de Cerámica y Vidrio (CSIC) of Madrid. The thickness of the Fe films and Fe/Au bilayers was controlled by the Q-microbalance. These depositions were performed at room temperature and under a 10^{-6} torr pressure, where the currents and deposition rates were varied depending on material and the type of deposition.

On the other hand, for the preparation of the CoPc/Au bilayers, CoPc films were evaporated from Cobalt(II) phthalocyanine ($C_{32}H_{16}CoN_8$) in powder, using a ceramic crucible. Before the CoPc evaporation, Au films were grown on substrates as indicated above. Prior to inserting the CoPcs into vacuum chamber to evaporate them, commercial CoPcs (from Sigma-Aldrich) were purified via sublimation in a vacuum furnace at 10^{-3} torr and 300 °C. The growth of CoPc thin films on Au films was carried out at 45 °C in two steps: first, a degasification of Pcs to remove possible impurities while the Au films are covered by a shutter; second, the evaporation of CoPcs under a 10^{-9} torr pressure where the shutter is opened and the molecules are evaporated on the Au films. Homogeneous deposition was achieved by a moderate rotation of substrate plate. CoPc films were grown at room temperature (α-CoPcs) or at 200 °C (β-CoPcs). For this work, CoPc/Au bilayers were fabricated by the homemade evaporation chamber in the Schuller Nanoscience Group, at the University of California, San Diego. The thickness of the CoPc/Au bilayers was controlled by the Q-microbalance and a shutter was used to cover the samples between the deposition of each layer (Au and CoPc layer).

---

[1] Au shows a crystalline structure face-centered cubic (FCC) with a lattice constant of 4.080 Å.

[2] Commonly, Fe is found in its body-centered cubic (BCC) crystalline structure, with a lattice constant of 3.515 Å.

2.2.2 Fabrication of Nanostructures by Annealing of Thin Films

Metallic thin films grown on substrates with poor adhesion are known to exhibit substantial modifications when annealing in air or vacuum, including the formation of hillocks or strain-free extrusions, nucleation and growth of holes and agglomeration of islands leading to a discrete structure [8–13]. The difference in thermal expansion coefficient between the metal and the substrate can lead to the formation of hillocks due to the relaxation of the thermal stress during the course of the annealing process. The formation of hillocks triggers the dewetting process, which is given by the minimization of surface and interface energies: the surface energy of the film and the substrate, and that of the film/substrate interface [14, 15]. The hillock formation depends on the film thickness: when increasing the film thickness, the density of hillocks decreases and their average diameter increases. The decrease of the hillock density can be explained by the increase of the grain size in thicker films, since the process of the grain-boundary diffusion is a controlling mechanism for the hillock formation [16]. Higher annealing temperature favors the nucleation and growth of holes by surface diffusion. For larger temperatures, the holes percolate leading to the formation of islands, which can coalesce during the annealing changing and increasing the average island size in order to reduce the surface energy. This last stage of the process is known as recrystallization for individual particles and/or coalescence if multiples particles join together [17]. An example of the process is illustrated in Fig. 2.6 for a 45 nm Au film annealed at different temperatures.

This method to obtain nanostructures depends mainly on the high surface energy anisotropy [15] and the final nanostructure morphology can be modified and altered by different parameters, such as the film initial thickness, the morphology of substrate [17, 18] and annealing conditions. The formation process of nanostructures is more likely under presence of oxygen due to the relaxation of thermal stress by surface diffusion [10, 19]. This phenomenology was initially considered as a problem in the fabrication of microelectronic devices [16], but it has been recently used successfully to obtain noble metal nanoparticles [12, 20].

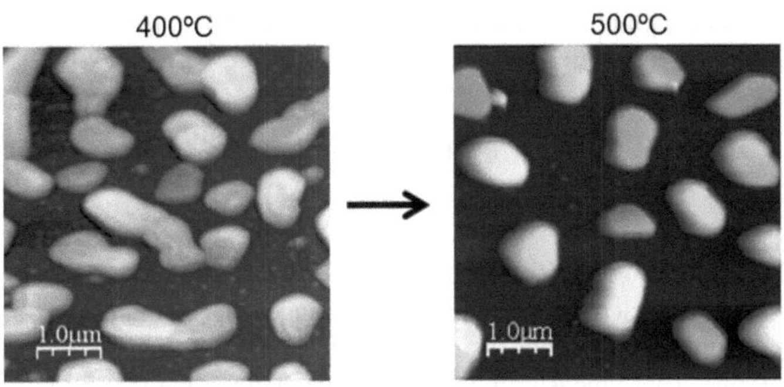

Fig. 2.6 Reshaping process for a 45 nm Au film annealed at 400 °C by recrystallization and coalescence when the sample is annealed at 500 °C

Table 2.1 Melting point and thermal expansion coefficient for different materials

Material	Melting point (°C)	Thermal expansion coefficient at 20 °C (10^{-6}/K)
Au	1063	14.2
Fe	1534	11.8
Soda-lime glass	700*	7.5
Silica glass	1600*	0.49

*Softening point

In this work, Au, FeO_x and Au/FeO_x nanostructures were fabricated by annealing of thin films and bilayers grown by thermal evaporation. The fabrication process, initial parameters (i.e., film thickness, substrate) and the annealing conditions (i.e., annealing time, temperature, atmosphere) for each kind of nanostructures are described later, in Chaps. 3 and 4. Table 2.1 shows the melting points and the thermal expansion coefficients [11, 21] of the materials and substrates used, which were taken into account for the fabrication of the different nanostructures.

2.3 Reflected Light Optical Microscopy (RLOM)

The first optical microscope is usually associated with *Zacharias Janssen*, in Middleburg, Holland, around the year 1595 [22]. No doubt that the invention of the optical microscope (as the invention of the telescope by *Galileo*) has signified a revolution in human knowledge. The optical microscope, as called light microscope, is an instrument that uses visible light to produce a magnified image of an object (or specimen) that can be observed by the eye or by an imaging device. Nowadays, optical microscopes can be very simple. However, they usually present many complex designs aimed to improve sample resolution and contrast. Two microscope components are of critical importance for forming the image. The condenser lens, which focuses light from the illuminator onto a small area of the specimen, and the objective lens, which collects light reflected by the specimen and forms a magnified real image at the real intermediate image plane near the eyepieces [23, 24]. The microscope resolution is limited by the diffraction limit and is defined as [25]:

$$d = \frac{0.61\lambda}{NA} \tag{2.1}$$

where d is the minimum resolved distance, λ is the wavelength and NA is the numerical aperture of the objective lens.

In this work, reflected optical microscopy images were obtained by a polarized light Zeiss Axiophot microscope, using the digital camera AxioCam MRc5 for taking the micrographs. These measurements were carried out at the Instituto de Cerámica y Vidrio (CSIC).

2.4 Scanning Electron Microscopy (SEM)

The electron microscope was developed by *Max Knoll* and *Ernst Ruska* in 1931 [26]. The scanning electron microscopy (SEM) is a type of electron microscopy that uses a focused beam of electrons to generate a variety of signals at the surface of samples. The types of signals that derive from electron-sample interactions include secondary electrons, backscattered electrons, X-rays, cathodoluminescence, specimen current and transmitted electrons, which reveal information about the sample (morphology, chemical composition, crystalline structure and orientation) [27]. Most of the work done on a SEM is for obtaining topographical information, which is mainly provided by secondary electrons.

The main components of an electronic microscope are: a vacuum system (it is required when using an electron beam because electrons will quickly disperse or scatter due to collisions with other molecules); an electron beam generation system (usually, a tungsten filament) that provides the beam of electrons known as the primary electron beam; an electron beam manipulation system that consists of electromagnetic lenses and coils that control the size, shape and position of the electron beam that reaches the sample surface; and the beam-specimen interaction system that involves the interaction of the electron beam with the sample and the types of signals that can be detected [28].

In this work, samples were studied by a TM-1000 Tabletop Microscope using Hitachi at the Instituto de Cerámica y Vidrio (CSIC). This electron microscope permits to obtain images without a special sample preparation, working with an accelerating voltage of 15 kV.

2.5 Atomic Force Microscopy (AFM)

Atomic force microscopy (AFM) was developed in 1986 by *Binning*, *Quate* and *Geber* [29]. AFM is a powerful tool that is used to study the morphology and structure of samples, allowing to characterize and manipulate a wide group of materials at the atomic level [30–32]. Easily, the AFM operation is based on measuring interactions between a tip and sample. Figure 2.7 shows the main interactions as a function of tip-sample distance.

The basic components of this system are a cantilever with a sharp tip (probe), a laser beam, a photodiode detector and a scanner, as illustrated in Fig. 2.8. The tip is attached to the free end of a cantilever and is approached very close to the surface of a sample. Attractive or repulsive forces, resulting from interactions between the tip and the surface, cause the bending of the cantilever. It is detected by means of a laser beam, which is reflected from the back side of the cantilever and collected in a photodiode. The variations on the topography of the sample will induce changes in the interactions between the tip and sample [30, 31], which will be detected by the bending of the cantilever. Figure 2.8 shows the basic concept of AFM, which can

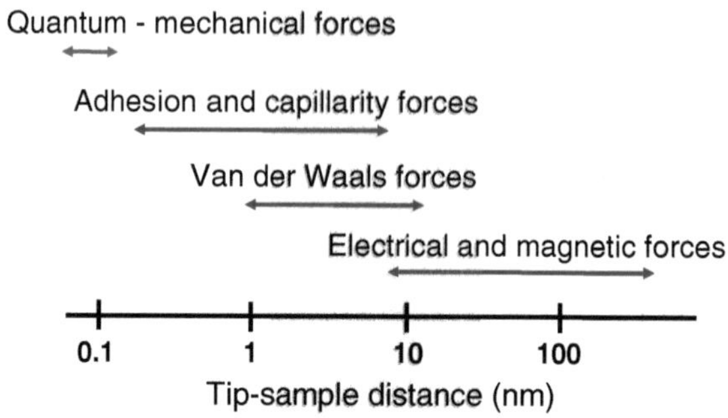

Fig. 2.7 Interaction forces between tip and sample in AFM

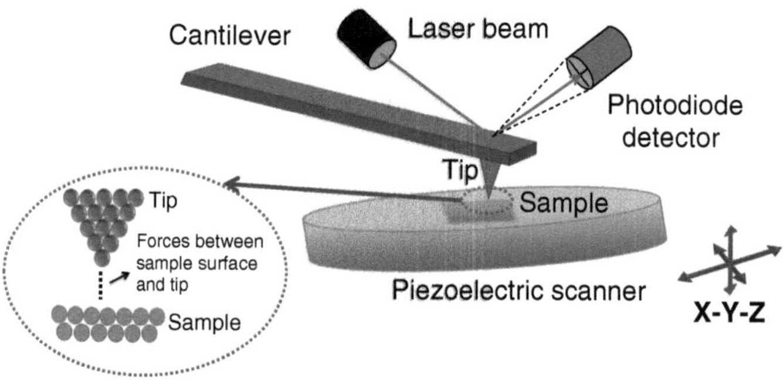

Fig. 2.8 Scheme of AFM

work in different modes of operation: contact mode, noncontact mode and tapping mode (intermittent contact mode).

The AFM can provide useful information through phase measurements, which collect the changes in the phase signal of the oscillating cantilever operating in noncontact or tapping mode of the microscope. Phase measurements are obtained simultaneously with the topography data and provide interesting information on localized tip-sample interactions on heterogeneous materials that present different surface properties, such as the composition, viscoelasticity or surface adhesion [33]. For materials with different properties, a shift in the phase signal between tip and sample is obtained, generating images with phase contrast [34], from which it is possible to distinguish the different local compositions.

In this work, topography and phase AFM images were taken with a Nanotec instrument of the Surface Science Group at the Universidad Complutense of Madrid. AFM measurements were performed in air, using silicon tips (with a resonant frequency of 200–300 kHz and an average tip radius $\leq$10 nm), in noncontact mode (the cantilever oscillates near its resonance frequency and the oscillation amplitude remains constant by a feedback system) and at room temperature. The images were analyzed using the *WSxM* software package from Nanotec [35].

2.6 Thickness Determination of the Films

The thickness of films was evaluated by different methods, depending on the vacuum chamber used to fabricate the samples:

- On the one hand, the thickness was determined using a quartz crystal microbalance, which was calibrated using X-ray reflectivity (XRR) performed on similar samples. The film thickness, in this case, is a nominal thickness. The chambers used in this way were: the home-made evaporation chamber in the Departamento de Electrocerámica at the Instituto de Cerámica y Vidrio of Madrid (CSIC) and the home-made evaporation chamber in the Schuller Nanoscience Group, at the University of California, San Diego.
- On the other hand, film thickness was evaluated using a linear relation obtained by the thickness (calculated using XRR) of similar films and their integrated absorbance in the range between 400 and 700 nm, obtained by the optical absorption spectra. Thicknesses calculated by this method correspond to Au films grown in the commercial Pfeiffer 306 auto coater, in the Departamento de Física de Materiales at the Universidad Complutense of Madrid.

2.6.1 X-Ray Reflectivity (XRR)

X-ray reflectivity (XRR) is a technique based on the X-ray diffraction phenomenon [36]. XRR provides information about the thickness, density and roughness of each layer of a thin film [37].

For the X-rays, the refractive index, n, of materials is slightly less than 1 and is given by [36, 37]:

$$n = 1 - \delta + i\beta \tag{2.2}$$

where δ and β represent the dispersion and the absorption, respectively. For incident angles larger than the critical angle, the X-rays penetrate inside material (see Fig. 2.9). In this way, reflection occurs at the top and the bottom surface of the film. By measuring the total reflection intensity (reflectivity) from grazing incident X-ray beam as a function of the incident angle with respect to the thin film surface, a XRR profile can be obtained.

The XRR profile shows oscillations caused by the X-ray interference that occurs between the X-rays reflected from the sample surface and the sample/substrate interface.[3] The oscillations of the XRR profile depend on the film thickness: the shorter period of the oscillations, the thicker films. From the reflectivity profile as a function of the incident angle, it is possible to determinate the film thickness using the following linear fit:

---

[3]For convectional θ–2θ measurements, we obtain information from the lattice planes that are parallel to the sample surface.

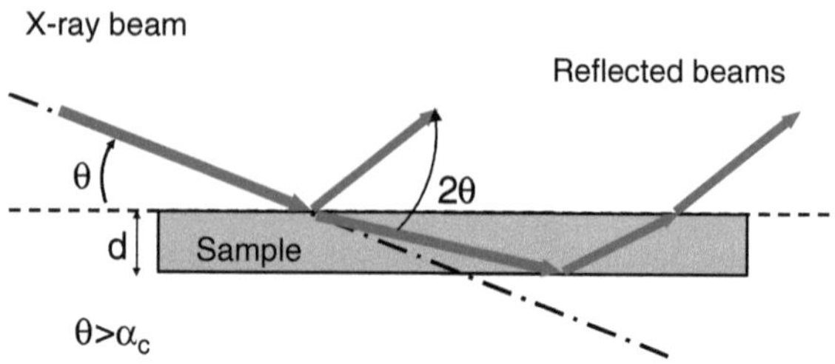

Fig. 2.9 Reflection and refraction of X-rays at material surface with the incident angle > total reflection critical angle

$$\alpha_m{}^2 = \alpha_c{}^2 + (m + \Delta m)^2 \frac{\lambda^2}{4d^2} \tag{2.3}$$

where m indicates the maximum or minimum order of the oscillations, Δm is 0 for minima or $\frac{1}{2}$ for maxima, λ is the wavelength of incident beam (in this work, $\lambda = 1.54$ Å for Cu-K$_\alpha$ X-rays), α_m corresponds to the angle of the maximum or minimum, α_c corresponds to the critical angle and d is the film thickness.

2.6.2 Relation Between the Thickness of Films and Their Integrated Absorbance

In order to know the thickness of thin films using a fast and accessible route, a linear relation between the thickness of Au thin films (just Au thin films were grown in the commercial Pfeiffer 306 auto coater) and their integrated absorbance, in the range between 400 and 700 nm, was determined. First, the thickness of a set of samples was obtained using XRR (see Fig. 2.10a), according to Eq. 2.3. Table 2.2 shows the thicknesses obtained by XRR for four Au films. Later, the integrated absorbance, obtained by the optical absorption spectra, was calculated in the range between 400 and 700 nm (range marked with arrows in Fig. 2.10b).

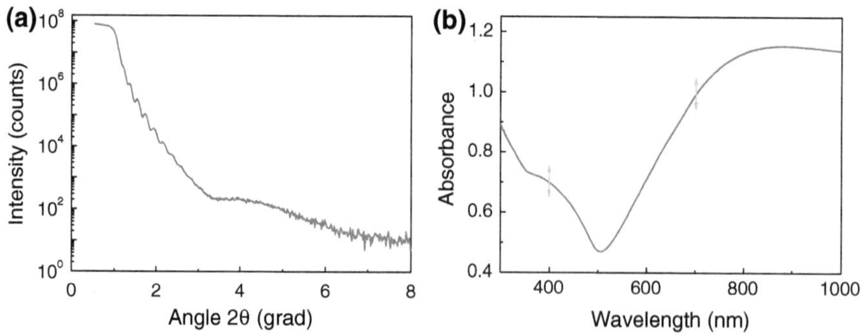

Fig. 2.10 **a** XRR curve and **b** optical absorption for a Au thin film with a width of 36 ± 3 nm

Table 2.2 Thickness obtained by XRR, according to Eq. 2.3, for four Au thin films

Sample	Thickness (nm)
1	36 ± 3
2	45 ± 1
3	54 ± 1
4	59 ± 1

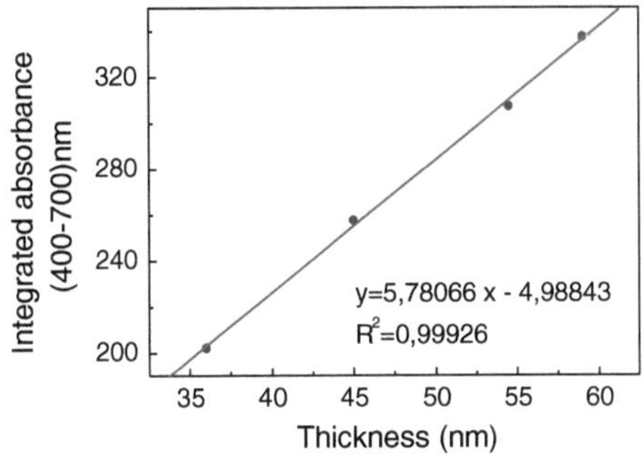

Fig. 2.11 Linear relation between the thickness and the integrated absorption in the range between 400 and 700 nm

Finally, representing the integrated absorbance in the range between 400 and 700 nm as a function of the Au thin film thickness obtained by XRR (see Fig. 2.11), we obtain the following linear relationship:

$$d(nm) = \frac{I.A. + 4.988}{5.781} \tag{2.4}$$

where d and $I.A.$ represent the thickness and the integrated absorbance, respectively.

In this way, the thickness of the rest of the Au films was evaluated by Eq. 2.4, measuring the optical absorption spectrum and calculating its integrated absorbance in the range between 400 and 700 nm.

2.7 Raman Spectroscopy

Raman spectroscopy is a non-destructive technique that studies the vibrational, rotational and other low frequency transitions of molecules or crystal atoms. Raman effect was discovered by C.V. Raman in 1928 [38] and it is based on the inelastic scattering of monochromatic light, usually from a laser source. It depends on the polarizability of the molecules and it requires a change in the frequency of photons from monochromatic light upon interaction with a sample [3].

When a light source (laser) is focused on a material, most of this energy will be elastically scattered. In this case, the molecules of the substance are excited to a virtual electronic state and immediately fall back to their original state by releasing a photon. If the photon energy of this scattered light is equal to that of the incoming light, the

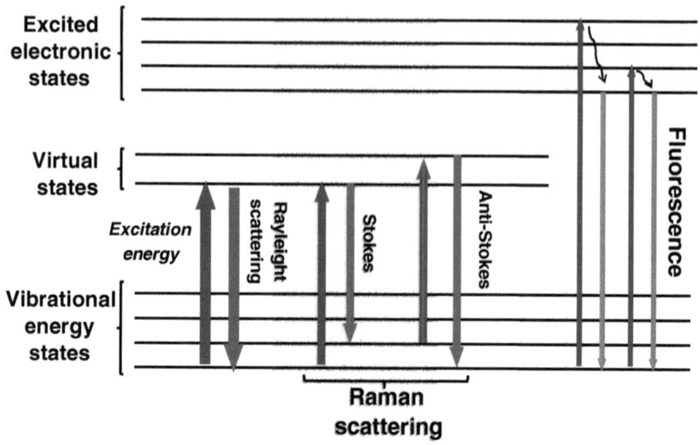

Fig. 2.12 Mechanisms of various light-scattering processes involved in Raman spectroscopy

process is called Rayleigh scattering. However, a molecule may also fall back from an excited electronic state to an energy state that is higher (Stokes type scattering, photon emission) or lower (anti-Stokes type scattering, photon absorption) than the original state (see Fig. 2.12). The difference of energy between the incoming and scattered photons (Raman shift) corresponds to the energy difference between vibrational energy levels of the molecule. The different vibrational modes of a molecule can, therefore, be identified by recognizing Raman shifts (or bands) in the inelastically scattered light spectrum [39].

Raman spectroscopy is extraordinarily sensitive to the chemical nature and the physical state of a sample. In the Raman spectra, the band position, intensity and bandwidth are closely related to the order, crystal size and defects in the samples [40]. The band position is sensitive to the presence of stresses or strains: a tensile stress will determine an increase in the lattice spacing and, hence, a decrease in the wavenumber of the vibrational mode. In the case of compressive strain, the decrease of the lattice parameter yields a corresponding increase of the vibrational frequency. The presence of crystalline disorder also produces changes in the frequency of the band, usually towards lower wavenumbers. These are related to the breaking of translational symmetry in the crystal, which can be due to structural defects, such as grain boundaries in nanocrystalline materials or dislocations. With respect to Raman bandwidth and bandshape, these are closely related to the crystalline order and the defect density. In principle, the bandwidth is related to the lifetime of the phonons. The presence of crystalline disorder produces a decrease of the phonon lifetime, generating an increase of the bandwidth. Other parameter very sensitive to the structure of crystals is the intensity of the Raman bands. The amount of material is directly related to the Raman intensity and a damage in the lattice leads to a decrease in the intensity of modes, related to the breaking of bonds and changes in atomic force displacements [40].

Raman spectroscopy is employed to distinguish different materials and to determine the kinds of species in a same sample. Material as the metals (i.e., Au) can lead to optical emissions as the fluorescence phenomenon that can mask the Raman spectra of other materials. Many efforts have been carried out in the last years to avoid the fluorescence or to separate the Raman signal of a material from fluorescence emission [41]. However, useful information can be obtained by the fluorescence phenomenon, as we show in Chap. 4.

In this work, samples were investigated by a confocal Raman microscope [42] using the Witec Micro-Raman Confocal (ALPHA 300R). Measurements were carried out at room temperature with a Nd:YAG laser (532 nm) in p-polarization. The optical resolution of the confocal microscope in the lateral direction is of 200 and of 500 nm in the vertical direction. The spectral resolution of the system is 0.02 cm^{-1}, under the best measurement conditions. Raman spectra were recorded in the spectral range 0–3600 cm^{-1}. Samples were mounted in a piezo-driven scan platform with a positioning accuracy of 4 nm in lateral and 0.5 nm in vertical. In order to prevent any damage or oxidation in the samples, laser excitation power was fixed at 0.7 mW. Raman measurements were performed using a objective with a lens of numerical aperture, NA, of 0.9. Collected spectra were analyzed by using *Witec* Control Plus Software.

2.8 X-Ray Absorption Spectroscopy (XAS)

X-ray absorption spectroscopy (XAS) is a technique that provides element-specific information about the local geometric and/or electronic structure of a material. The experiments can be carried out at synchrotron radiation sources, which provide intense and tunable X-ray beams.

When the X-rays reach a material, the intensity of electromagnetic radiation is reduced by different processes of interaction, such as scattering or absorption (see Fig. 2.13). For X-ray photons, with energies ranging from 100 to 500 keV, the photoelectric effect is dominating, where an X-ray is absorbed and a core level electron is promoted out of the atom [43]. Hard X-rays are termed so when the energy is of the order of tens of keV, while those with lower energy are called soft X-rays [44]. With soft X-rays, excited levels are close to Fermi level and so they are more sensitive to electronic structure. However, hard X-rays are more sensitive to crystallographic structures, since the excited levels are more internal. When the monochromatic X-ray beam of intensity I_0, passes through a sample of thickness x, the intensity is reduced according to the Lambert-Beer law:

$$I = I_0 e^{-\mu x} \tag{2.5}$$

where I is the transmitted intensity and μ represents the linear absorption coefficient, which depends on the types of atoms and the material density and the energy.

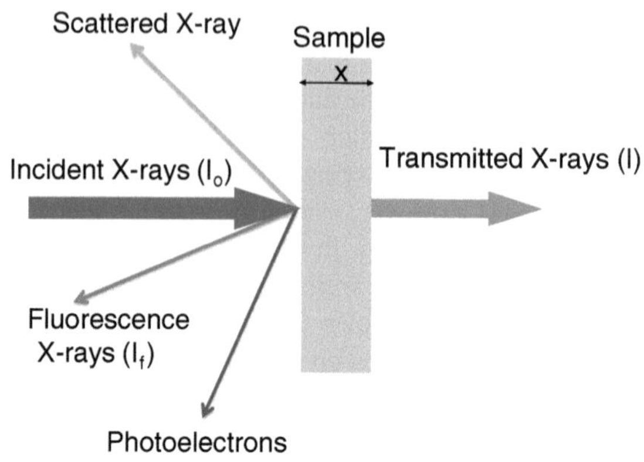

Fig. 2.13 X-ray absorption measurements: X-ray interaction with a sample

At certain energies, due to the discrete character of electronic levels, the absorption increases drastically and gives rise to an absorption edge. Such edge occurs when the energy of the incident photons, E, is just sufficient to cause excitation of a core electron of the absorbing atom to a continuum state, producing a photoelectron. In this process, the energies of the absorbed radiation at these edges correspond to the binding energies of electrons in the K, L, M, etc., shells of the absorbing elements.

An X-ray absorption spectrum (see Fig. 2.14) is generally divided into four sections [43, 45]:

- *Pre-edge*: where the $E < E_0$, being the E_0 the edge energy. The pre-edge together the post-edge are used for normalization of XAS spectra.
- *Absorption edge*: rapid increase of absorption at E_0, which corresponds to the electron excitation energy from core levels to Fermi level. This region is very sensitive to the oxidation state of the materials, coordination number and it gives information about the electronic structure of the atom [43, 45]. The absorption edge is used for calibration of XAS spectra.
- *X-ray absorption near edge structure (XANES) region*: where the energy of the incident X-ray beam, E, is around 10 eV before and up to 50 eV above the edge.

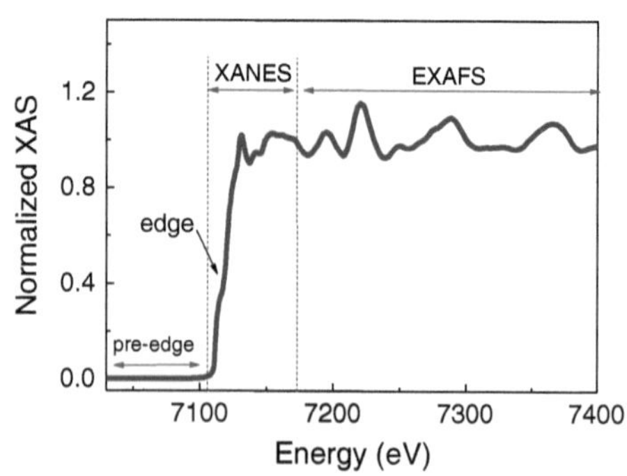

Fig. 2.14 XAS spectrum at Fe K edge measured in transmission mode for an Fe foil, at branch A of the BM25 SpLine beamline at the ESRF

In the XANES region, transitions occur from core electrons to non-bound levels with energy close to Fermi energy. This region shows narrow and intense peaks because of the high probability of such transitions.

- *Extended X-ray absorption fine structure (EXAFS) region*, which is approximately from 50 eV above absorption edge and continues up to 1000 eV above the edge. This region shows soft oscillations of absorption coefficient due to the photo-electron scattering with the neighbor atoms. Studying EXAFS region, interatomic distances can be resolved as well as the identification of neighbor atoms.

In this work, XAS measurements were performed at the branch A of the CRG BM25-SpLine Beamline at the European Synchrotron Radiation Facility (ESRF) in Grenoble (France) [46, 47]. This beamline is split into two branches, A and B, each of them fully equipped with focusing optics and experimental stations that are independently operated.

Branch A enables the performance of XAS and high resolution powder diffraction measurements. This branch is located on the soft edge of the D25 bending magnet with a critical energy of 9.7 keV and energy resolution of $\Delta E/E = 1.5 \times 10^{-4}$. The X-ray energy ranges between 5 and 45 keV and the flux is of the order of 10^{12} photons/s at 200 mA ring current. The beam spot size can be varied for the whole energy range between $300 \times 100\ \mu m^2$ and $40 \times 10\ mm^2$ in the horizontal and vertical directions, respectively. The position and dimensions of the focused beam are kept constant during a $\sim$1 keV energy scan, which represents standard conditions for EXAFS measurements. The XAS system is arranged onto an optical table for optimum placement and alignment of the environmental sample and the detection equipment components. The system is equipped with motors covering all degrees of freedom, three translation stages (X, Y, Z) and three rotation stages (θ, χ, φ) for centering the sample. The precision of such rocking cradle motors is 0.001° and they can tilt the sample stage within a range of ±15° along the directions parallel and perpendicular to the incident beam. While the beamline can operate in both transmission and fluorescence modes, only the later is used in this work for studying the samples. A gas ionization chamber (from Oken) working in the low-pressure range is placed for monitoring the incoming beam intensity. Fluorescence is detected with a nitrogen-cooled 13-element Si(Li) detector (from e2v Scientific Instruments). The low temperature of the Si(Li) and FET ensemble also diminishes leakage current and electronic noise. The Si(Li) fluorescence detector allows performing fluorescence measurements in a range between 3 and 30 keV. The energy resolution at 6 μs peaking time is around 140 eV in every crystal. The detector is also fully equipped with translation and rotation motors to optimize its position, reducing in this way the solid angle for elastic diffuse scattering contributions.

In this work, XAS measurements were performed at Au L edge (E = 11.919 keV), Fe K edge (7.112 keV) and Co K edge (E = 7.720 keV) at room temperature. Besides measuring XAS spectra, the X-rays were used to induce modifications on different samples [48]. The measurements were analyzed with *ATHENA* software [49].

2.9 Optical Absorption Spectroscopy

In a spectrophotometer of optical absorption, a monochromatic light source that is varied in a wide range of wavelengths passes through a material, promoting an electron from its fundamental state to a state of higher energy. In the process, electronic and vibrational transitions of material and their impurities, defects, etc., absorb light at a certain energy, showing zones in the absorption spectrum with high intensity, which are characteristics of each material [3]. Absorption bands corresponding to localized surface plasmons (LSPs) can be also determined by the optical absorption spectroscopy.

The absorbance (A) of a material can be defined as the logarithm of the ratio between incident and absorbed light intensity [50]:

$$A = \log \frac{I_0}{I} = -\log T \tag{2.6}$$

where I_0 represents the incident intensity, I represents the transmitted intensity and T is the transmittance. The absorbance is adimensional [50].

In this work, the optical absorption bands of different samples were studied using two double-beam ultraviolet-visible spectrophotometers. A scheme of a double-beam spectrophotometer is shown in Fig. 2.15, where the beam from the light source is split in two: one beam illuminates the sample and the other one illuminates the reference (i.e., a substrate). Afterwards, the intensity of both beams is compared and the transmittance or absorbance of the sample can be obtained. Optical absorption spectra were recorded with:

- A Shimadzu UV-1603 double beam spectrophotometer in the Departamento de Física de Materiales at the Universidad Complutense of Madrid.
- A V-670 UV-visible double beam spectrophotometer in the Dipartimento di Chimica at the Universitá degli Studi di Firenze.

The optical absorption measurements were carried out in transmission mode and at room temperature. Baseline correction procedure (a spectrum in air was taken as baseline) was executed prior to each measurement session.

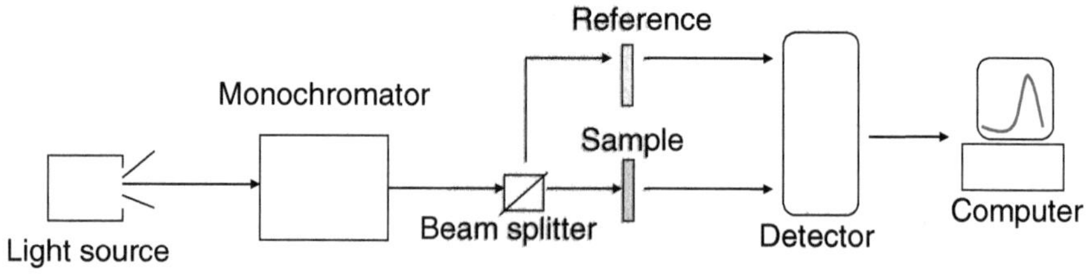

Fig. 2.15 Scheme of a double-beam ultraviolet-visible spectrophotometer

2.10 Surface Plasmon Resonance (SPR) in the Kretschmann-Raether Configuration

The Kretschmann-Raether configuration [1, 51, 52] is mostly employed for the excitation of surface plasmons (SPs) in thin films. Figure 2.16 displays a scheme of the home-made surface plasmon resonance (SPR) device designed and mounted in the Departamento de Electrocerámica at the Instituto de Cerámica y Vidrio (CSIC) of Madrid, showing the different elements of the experimental setup. Figure 2.17 shows a picture of the system.

In this geometry, SPs are excited in the attenuated total reflectance mode using a HeNe (632.8 nm) linearly polarized laser in a p-polarization geometry, with respect to the sample surface (the electric field in the plane of incident light). The laser is mounted on a cradle equipped with yaw and pitch movements for an accurate orientation of the beam on the sample. A linear polarizer is used to have a well-defined p-polarization. Subsequently, a beam splitter [53] deflects the laser intensity to a photodetector, in order to record fluctuations in laser intensity during the experiments. These fluctuations are of the order of 3 %, decreasing with time. Beyond the beam splitter, the laser beam is modulated with an optical chopper, working at 479 Hz. Finally, the laser light reaches the sample, which is coupled to a triangular quartz prism (semicircular prism can be also used). The sample consists of a thin metallic film (typically 50 nm of Au or Ag, using in this work Au films) grown on a glass substrate. Overlayers of dielectric materials can be grown. Sample is fixed to the prism thought the substrate side using gel index matching for a good coupling. Both (sample and prism) are mounted on top of a rotating motor that allows varying the laser incidence angle (see Figs. 2.16 and 2.17). The sample stage has an independent XYZ translation stage for sample positioning. During the measurements, the laser beam reflected at the sample is collected by an elongated photodiode (to avoid moving the detector during the angle scan) as a function of the incident angle, by rotating

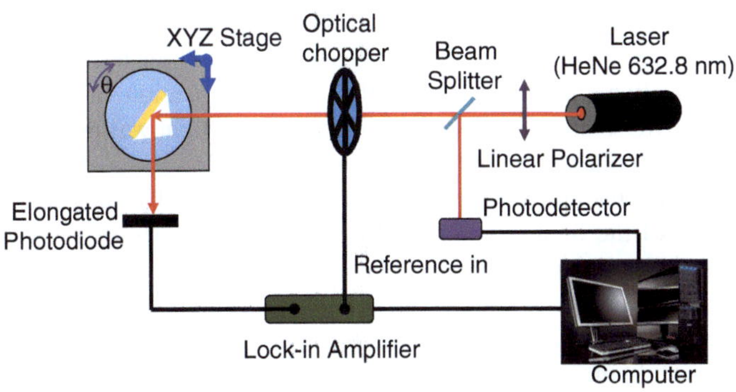

Fig. 2.16 A scheme of SPR setup in the Kretschmann-Raether configuration

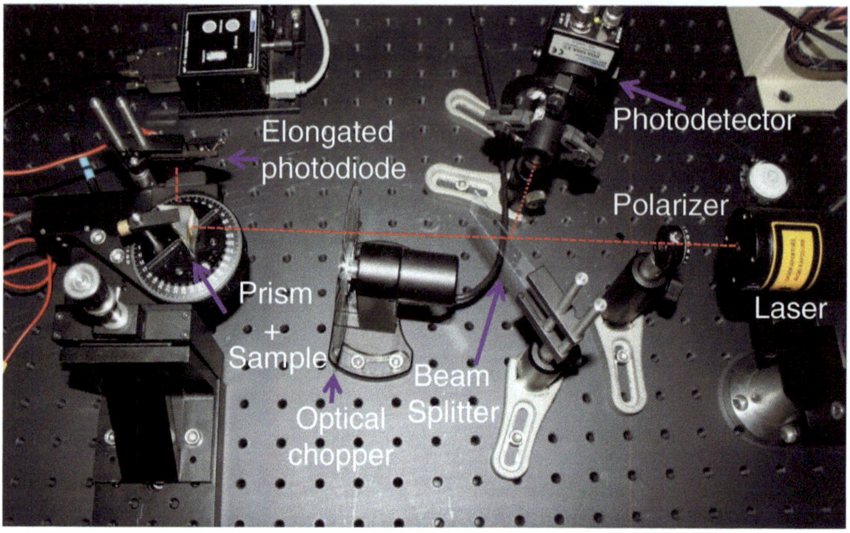

Fig. 2.17 Photograph of the mounted SPR device in the Departamento de Electrocerámica at the Instituto de Cerámica y Vidrio (CSIC) of Madrid, remarking the different elements

the sample [1]. The photodiode signal is registered with a lock-in amplifier, using the optical chopper frequency as reference frequency.

A detailed description of different elements employed in this experimental setup is enumerated below:

- HeNe laser, 632.8 nm: a Thorlabs linearly polarized laser (power 0.8 mW) in a p-polarization geometry with respect to the sample surface was used. The laser spot diameter is 0.48 mm, measured as the full width half maximum (FWHM) of Gaussian profile of the spot. The coherence length is around 30 cm.
- Beam splitter: a beam splitter optical device (microscope slide) was used to split the light beam in two. The beam splitter deflects about 5 % of the laser intensity.
- Linear polarizer: this is an optical filter that can convert a light beam of undefined or mixed polarization into a beam with well-defined polarization. In the experimental setup, in the Kretschmann-Raether configuration, the light was p-polarized with a Thorlabs linear polarizer, in order to excite the SPs [1].
- Optical chopper: the MC2000 optical chopper from Thorlabs was used to modulate the intensity of the light beam. Usually, it is a rotating slitted mechanical disc through which the laser source passes before reaching the sample. The laser beam was modulated at 479 Hz.
- Photodetector: a photodetector operates for converting the incident light to a voltage or current. In the experimental setup, a DET10A Thorlabs photodetector was used to record fluctuations of laser intensity when it is split by the beam splitter (laser intensity reference).

- Elongated Photodiode [54]: the most common used photodetector is the photodiode. In this device, the S2387-130R Hamamatsu Si photodiode was used working in photoconductive mode (there are two modes of operating: photovoltaic and photoconductive modes) with a load resistance of 10 kΩ and applying a voltage of 8 V. In this mode, the photodiode is reverse biased, obtaining a current proportional to the light intensity. In the SPR measurements, the photodiode signal depends on both the light intensity and the incidence angle, and therefore the photodiode sensitivity changes along the SPR curve. Consequently, the measured SPR spectrum can be distorted, affecting the fits and numerical analysis of SPR curves [55]. To avoid this, SPR measurements were carried out under the same conditions, with the photodiode placed in the same position, parallel to the laser beam.
- Lock-in amplifier [56]: the lock-in amplifier or phase-sensitive detector, SRS Model SR830, was used to collect the photodiode signal. Usually, a lock-in amplifier is employed to improve the signal-to-noise ratio. These devices are able to isolate the signal of a specific frequency (i.e., reference frequency of chopper), so that noise signals at different frequencies than that of reference are rejected and they do not affect the measurement.

In the Kretschmann-Raether configuration, using triangular prisms, only the points along the rotation axis remain at the same position when the sample rotates. If the laser spot does not impinge at the sample on this rotation axis, its position will change during the angle scan. This is not critical when samples are homogeneous. To avoid misalignments, semicircular prisms are preferable since for triangular prisms the refraction at the prism surface will induce a certain horizontal deviation of the laser beam, which is dependent on the incident angle. For instance, for 1 cm side triangular quartz (n = 1.457 at $\lambda = 633$ nm [7]) prism, when scanning the incidence angle in a range of 10°, the spot will move about 0.6 mm. However, the cross-section of a semicircular prism is not completely semicircular and this must be reduced by the thickness of the glass substrate, using glass substrates with a specific thickness. Such limitation on the thickness of glass substrate does not exit with triangular prism [57]. Besides this, using semicircular prisms with a small width comparable with the spot size of the laser, focalization problems can be obtained due to the wavefront is no plane. For this reason, in this work, triangular quartz prisms are used and measurements were carried out with the laser spot impinges at the sample on this rotation axis.

As previously discussed in Chap. 1, in the Kretschmann-Raether configuration, the glass/metal interface is illuminated with the laser beam in total reflection conditions. In these conditions, the evanescent field of the incident light propagates though the metallic film looking for the metallic/dielectric interface. For a certain incidence angle, the dispersion relationship is coincident with that of the SPs and these are excited. In this way, the reflectivity versus incident angle can be measured, obtaining the SPR spectrum. Figure 2.18 shows a SPR spectrum for a 50 nm Au film. In the SPR spectrum, we can distinguish the resonance angle (θ_R), which corresponds to

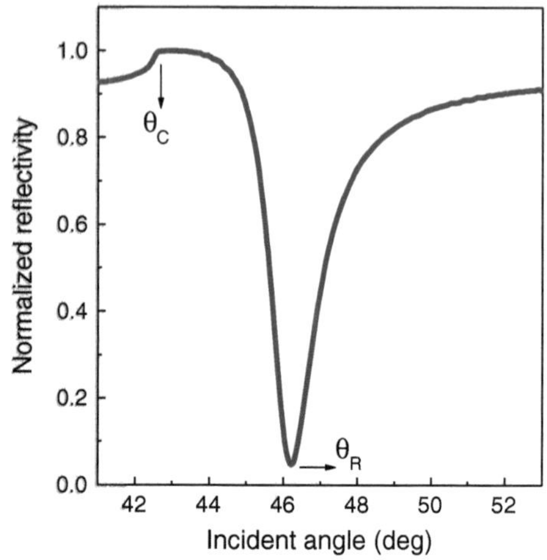

Fig. 2.18 SPR spectrum for a 50 nm Au film grown on a silica substrate

the incident angle for which the reflected intensity shows a minimum and the critical angle (θ_C) that is the incident angle above which the total internal reflection occurs [1] (see Fig. 2.18). The resonance angle and the whole SPR spectrum depend on different parameters, such as dielectric constants and the thickness of both the metallic thin film and its surrounding media (substrate and possible overlayers). However, the critical angle depends just on the refractive index of the quartz prism and the air. The total reflection takes place at the substrate/Au interface. When the laser passes from the prism to the substrate, according to Snell's law, the refraction keeps the component of the light wavevector parallel to the surface in both media constant. As the substrate/Au interface is illuminated in total reflection conditions, there is an evanescent field propagating through the second medium that does not suffer refraction but keeps the wavevector of the incident medium. The main limitation and advantage of the evanescent field is that it decays exponentially in the perpendicular direction to the interface, so its effects are limited to a few nanometers (see Fig. 1.14 in Chap. 1).

As the critical angle depends just on the refractive index of the quartz prism and the air, and these remain fixed in this work, we use this angle as parameter to adjust and normalize the SPR spectra. A slight drift of the rotating motor, on which sample and prism are mounted, was observed. In our measurements, this drift is corrected by adjusting the θ_C of the SPR spectra to a same value: using a quart prism with a refractive index of 1.457 for $\lambda = 633$ nm and assuming the air refractive index as 1, θ_C is 42.61°, taking into account the prism refractions.

In this work, this home-made device was used to measure and analyze the extended surface plasmon resonance (ESPR) in Au thin films, in order to study the characteristics of both the metallic film and the surrounding media.

2.10.1 Software to Measure SPR Spectra

Data of the laser beam reflected as a function of the incident angle were collected with a computer using a home-made VISUAL BASIC code. This software allows observing the SPR spectra while the measurements are performed (see Fig. 2.19). Besides this, it is possible to measure the laser intensity deflected by the beam splitter to a photodetector during the experiments. Therefore, we can examine both the SPR reflectivity and the same normalized by the laser intensity deflected (reference) as a function of the incident angle. The software allows the modification of measurement parameters such as the step per point, the integrating time in each measurement and the range of incident angle. This software generates files in the ASCII code for the data processing.

2.10.2 Reproducibility of SPR Spectra

Figure 2.20a shows five consecutive SPR spectra for a 52 nm Au film grown onto a glass substrate, which have been collected in the same conditions. The relative

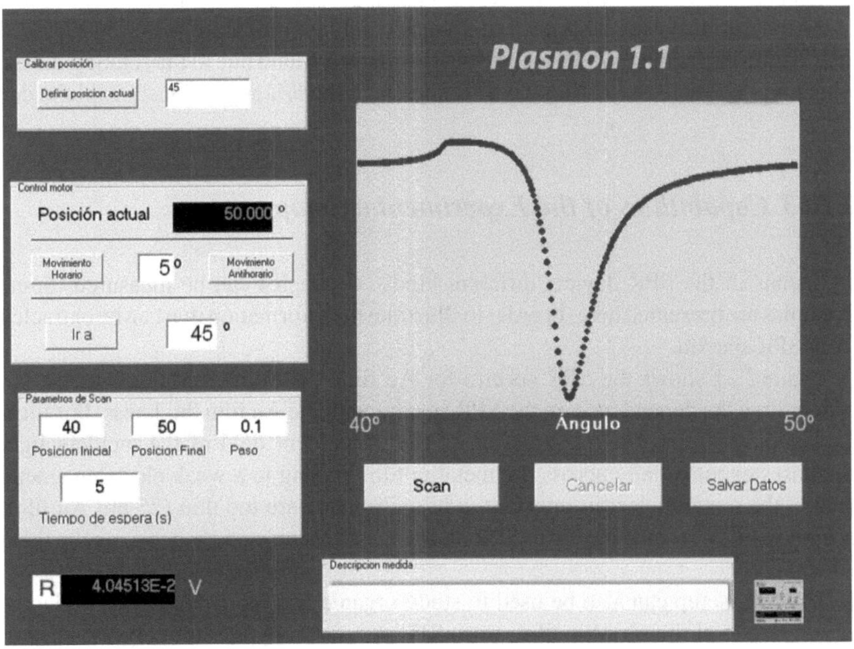

Fig. 2.19 SPR measurement with the home-made VISUAL BASIC code

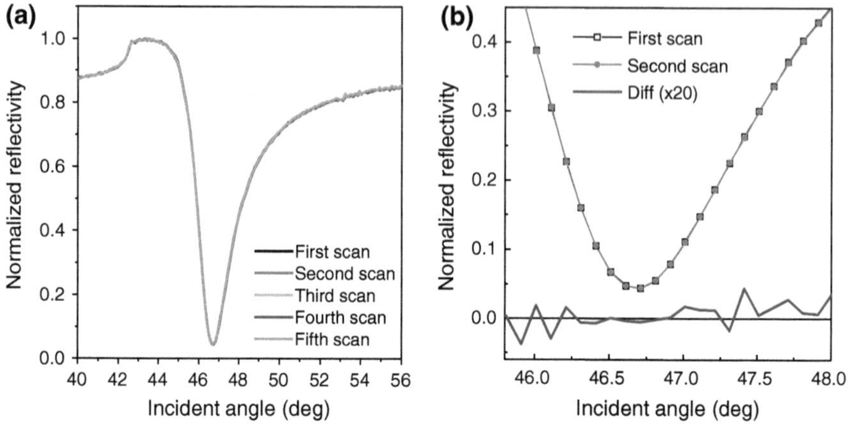

Fig. 2.20 **a** Consecutive spectra for a 52 nm Au film grown onto a glass substrate obtained with the SPR device shown in Fig. 2.16 and **b** a detail of the resonance region between the first and second scans and the difference in the scans (multiplied ×20)

difference between the first and second spectra (see Fig. 2.20b) is of the order of $\Delta R/R = 10^{-2}$ for the whole region of the spectrum. Therefore, with this equipment, one can clearly detect relative differences below 1 % with single scans. The resolution of the measurement can be improved upon accumulation of scans (scans in Fig. 2.20 take 20 min each one). A drift of the SPR spectra was found due to the rotating motor, which was corrected adjusting the SPR spectra at the θ_C, as previously explained.

2.10.3 Capabilities of the Experimental Setup

By means of the SPR device, different kinds of samples can be measured. Some examples are presented here, in order to illustrate the information that can be extracted from SPR spectra.

Figure 2.21 shows the SPR spectra for Au films with different thicknesses. We can observe the dependence on the SPR spectra with the Au film thickness. For films thicker than 80 nm, the transmission of the evanescent field of the incident light decreases exponentially across the metallic film leading to a weak electromagnetic field at the metal/dielectric interface, while films that are too thin (25 nm Au film) exhibit reduced resonance due to SPR damping and there is no chance to excite them effectively.

The SPR setup can also be used to study organic and inorganic dielectric layers grown on top of the metallic film, since SPR spectroscopy is extremely sensitive to the features of the dielectric medium over the metallic film [1, 58, 59]. Figure 2.22a shows the SPR spectra for a 2 and 5 nm Co-phthalocyanine (CoPc) layer grown at RT on a 50 nm Au film deposited onto a soda-lime substrate. We can observe the

Fig. 2.21 SPR spectra for Au films with different thickness: 25, 45, 70 and 90 nm

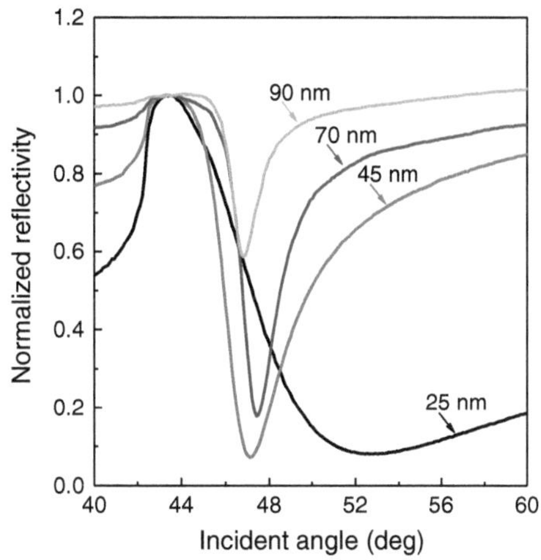

changes of the SPR spectrum depending on the CoPc layer thickness: the larger the CoPc thickness, the larger the decrease of the intensity and the larger is the shift of the θ_R towards larger angles. Figure 2.22b shows the SPR spectra for a 5 nm CoPc layer grown at RT and 200 °C on a 50 nm Au film deposited onto a soda-lime substrate. The sample grown at 200 °C shows a shift towards smaller angles and a reduction in the reflectivity at the resonance, with respect to the sample grown at RT with the same thickness. Literature reports that the growth temperature changes the morphology and grain size of CoPc layers [60], modifying their dielectric permittivity. These modifications of SPR curves can be studied to determinate the changes in both real and imaginary parts of the refractive index of CoPc layers with great accuracy [61].

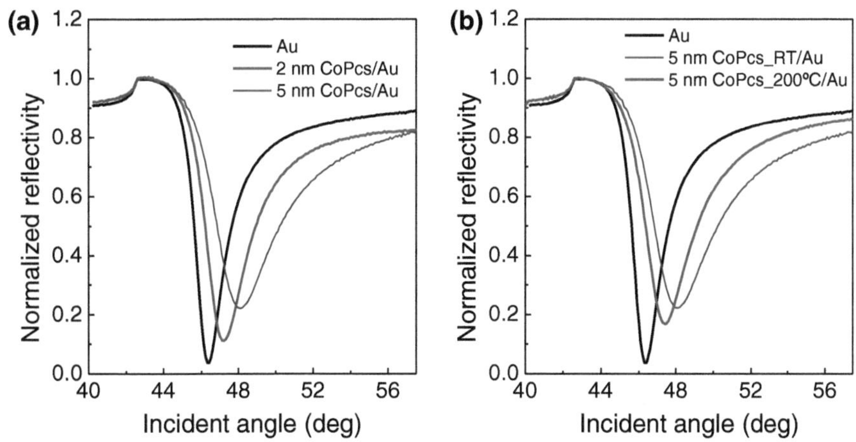

Fig. 2.22 SPR curves of **a** CoPc layers of 2 (*red line*) and 5 nm (*blue line*) nominal thicknesses grown at RT on a 50 nm Au film/soda-lime system and **b** CoPc layers of 5 nm nominal thickness grown at RT (*blue line*) and 200 °C (*red line*) on 50 nm Au film/soda-lime system

2.10.4 Software to Simulate SPR Spectra

Simulations of SPR spectra were performed with the software Winspall [62] version 3.02, a freeware software to calculate and simulate SPR spectra with great accuracy. This code is based on the Fresnel equations, including the correction of both reflection and refraction of the coupling prism. The software allows taking into consideration parameters such as the kind of prism (i.e., triangular, semicircular), the excitation light polarization or wavelength, the spectral range and the number of points.

Given a system, the calculation of SPR curves can be performed knowing the parameters (thickness and dielectric constants) of each layer. The layers are the different media of one system. For the case of a Au film grown on a silica substrate, we have a four-media system: prism, silica substrate, Au film and air. If a dielectric overlayer is grown on the metallic film, then a five-media system should be considered. On the other hand, simulations of SPR spectra can be carried out for different systems. For that, we can fix the known parameters and leave free the unknown ones. Manual and iterative simulations can be achieved. A manual simulation is considered good by visual inspection while the iterative one shows a divergence, which represents the deviation between the simulated and measured spectra.

Figure 2.23 illustrates the calculated SPR spectra for Au films with different thickness [62], showing the dependence of the resonance curve with the Au film thickness. Parameters such as the resonance angle, reflected intensity and the bandwidth depend on the film thickness. That can be observed in Fig. 2.24a, which shows the variation of the reflectivity at θ_R as a function of θ_R for Au films with different thicknesses, from calculated SPR curves. We find that the reflectivity at θ_R decreases when the film thickness increases up to a certain thickness, around 50 nm, where the reflected intensity increases. Figure 2.24b illustrates the decrease of the θ_R when the Au film thickness increases up to a thickness of around 50 nm, from which the θ_R remains almost constant.

In this work, the software Winspall is employed to calculate SPR spectra for different systems knowing the parameters of each layer (dielectric constants and thickness of each media). Simulations of measured SPR spectra are carried out for

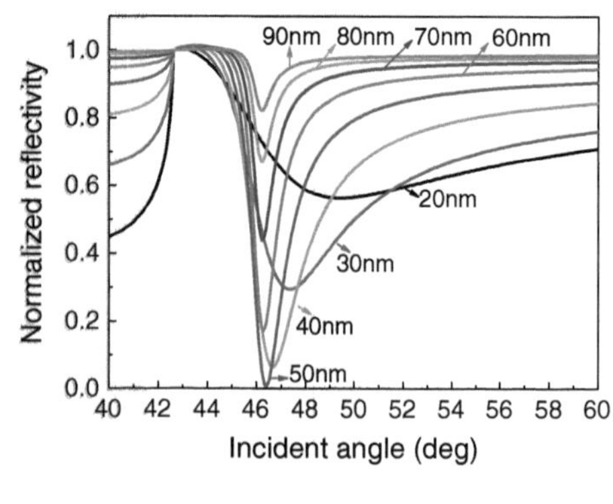

Fig. 2.23 Calculated SPR spectra for Au films ($\varepsilon_1 = -11.74$ and $\varepsilon_2 = 1.25$, according to Johnson and Christy [63]) with different thickness onto silica glass (n = 1.457 at $\lambda = 633$ nm)

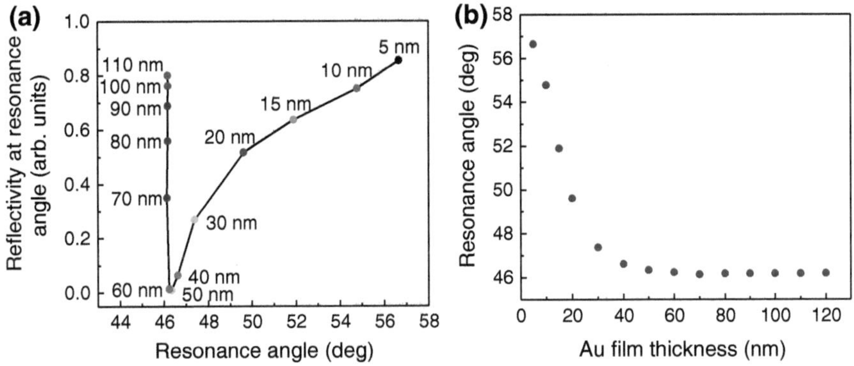

Fig. 2.24 **a** SP reflectivity as a function of the θ_R and **b** θ_R as a function of the thickness for Au films with different thicknesses. Data obtained from calculated SPR spectra

systems that show variations, in order to determine the modified layer and the new parameters [64] after modification. For each analyzed system, we show the kind of simulation, the free and fixed parameters, the values of the dielectric constants and the thickness of each layer.

References

1. H. Raether, *Surface Plasmons on Smooth and Rouge Surfaces and on Gratings* (Springer, Berlin, 1988)
2. J. Thornton, *Physical Vapor Deposition, Semiconductor Materials and Process Technology Handbook* (Noyes Publications, New Jersey, 1988)
3. J.M. Albella (ed.), *Láminas delgadas y recubrimientos: Preparación, propiedades y aplicaciones* (Instituto de Ciencia de Materiales, CSIC, Madrid, 2003)
4. K.S. Sreeharsha, *Principles of Physical Vapor Deposition of Thin Films* (Elsevier Science Ltd, New York, 2006)
5. E. Machlin, *Material, Science in Electronics: The Relationship between Thin Films Processing and Microstructure* (Giro Press, New York, 1995)
6. See http://www.valleydesign.com/sodalime.htm
7. See http://www.uqgoptics.com/
8. A.E.B. Presland, G.L. Price, D.L. Trimm, Hillock formation by surface diffusion on thin silver films. Surf. Sci. **29**(2), 424–434 (1972)
9. M.S. Rahman Khan, Hillock and island formation during annealing of gold films. Bull. Mater. Sci. **9**, 55–60 (1987)
10. S.K. Sharma, J. Spitz, Hillock formation, hole growth and agglomeration in thin silver films. Thin Solid Films **65**(3), 339–350 (1980)
11. J.A. Thornton, D.W. Hoffman, Stress-related effects in thin films. Thin Solid Films **171**(1), 5–31 (1989)
12. I. Doron-Mor, Z. Barkay, N. Filip-Granit, A. Vaskevich, I. Rubinstein, Ultrathin gold island films on silanized glass. Morphology and optical properties. Chem. Mater. **16**(18), 3476–3483 (2004)
13. X. Sun, H. Li, Gold nanoisland arrays by repeated deposition and post-deposition annealing for surface-enhanced Raman spectroscopy, Nanotechnology **24**(35) (2013)

14. M.S. Jackson, C.-Y. Li, Stress relaxation and hillock growth in thin films. Acta Metallurgica **30**(11), 1993–2000 (1982)
15. C.V. Thompson, Solid-state dewetting of thin films. Annu. Rev. Mater. Res. **42**(1), 399–434 (2012)
16. S.J. Hwang, J.H. Lee, C.O. Jeong, Y.C. Joo, Effect of film thickness and annealing temperature on hillock distributions in pure Al films. Scripta Materialia **56**(1), 17–20 (2007)
17. T. Tristan Leigh, Optical properties of metal nanoparticles and their influence on silicon solar cells. PhD thesis, University of Southamptom, 2009
18. P. Lansåker, P. Petersson, G. Niklasson, C. Granqvist, Thin sputter deposited gold films on In_2O_3:Sn, SnO_2:In, TiO_2 and glass: optical, electrical and structural effects. Sol. Energy Mater. Sol. Cells **117**, 462–470 (2013)
19. C. Yan, Y. Chen, A. Jin, M. Wang, X. Kong, X. Zhang, Y. Ju, L. Han, Molecule oxygen-driven shaping of gold islands under thermal annealing. Appl. Surf. Sci. **258**(1), 377–381 (2011)
20. S. Pillai, K.R. Catchpole, T. Trupke, M.A. Green, Surface plasmon enhanced silicon solar cells. J. Appl. Phys. **101**(9), 093105 (2007)
21. Y.S. Touloukian, R.K. Kirby, R.E. Taylor, P.D. Desai, Thermal Expansion: Metallic Elements and Alloys (Thermophysical Properties of Matter, vol 12), 1st edn. (IFI/Plenum, New York, 1975)
22. See http://www.history-of-the-microscope.org/hans-and-zacharias-jansen-microscope-history.php
23. S. Bradbury, B. Bracegirdle, *Introduction to Light Microscopy* (BIOS Scientific Publishers Ltd., Oxford, 1998)
24. B. Herman, J.J. Lemasters (eds.), *Optical Microscopy: Emerging Methods and Applications* (Academic Press, New York, 1993)
25. D.B. Murphy (ed.), *Fundamentals of Light Microscopy and Electronic Imaging* (Wiley, New York, 2001)
26. See http://inventors.about.com/od/mstartinventions/a/microscope_2.htm
27. See http://serc.carleton.edu/research_education/geochemsheets/techniques/SEM.html
28. M. Dunlap, D.J.E. Adaskaveg, *Introduction to the Scanning Electron Microscope. Theory, Practice and Procedures*. Facility for advanced instrumentation,1997
29. G. Binnig, C.F. Quate, C. Gerber, Atomic force microscope. Phys. Rev. Lett. **56**, 930–933 (1986)
30. V.N. Paredes, *Estudio de las propiedades mecánicas de superficies de Au y TiO_2 mediante microscopio de fuerzas*. PhD thesis, Universidad Complutense de Madrid, Febrero 2009
31. F.J. Giessibl, Advances in atomic force microscopy. Rev. Mod. Phys. **75**, 949–983 (2003)
32. N. Jalili, K. Laxminarayana, A review of atomic force microscopy imaging systems: application to molecular metrology and biological sciences. Mechatronics **14**(8), 907–945 (2004)
33. B. Bhushan, J. Qi, Phase contrast imaging of nanocomposites and molecularly thick lubricant films in magnetic media. Nanotechnology **14**(8), 886 (2003)
34. R. Brandsch, G. Bar, M.-H. Whangbo, On the factors affecting the contrast of height and phase images in tapping mode atomic force microscopy. Langmuir **13**(24), 6349–6353 (1997)
35. I. Horcas, R. Fernández, J.M. Gómez-Rodríguez, J. Colchero, J. Gómez-Herrero, A.M. Baro, $WSXM$: a software for scanning probe microscopy and a tool for nanotechnology. Rev. Sci. Instr. **78**(1), 013705 (2007)
36. K. Inaba, X-ray thin-film measurement techniques. Overview. Rigaku J. **24**(1), (2008)
37. M. Yasaka, X-ray thin-film measurement techniques. X-ray reflectivity measurement. Rigaku J. **26**(2) (2010)
38. C.V. Raman, A new radiation. Indian J. Phys. **2**, 387–398 (1928)
39. J. Ferraro, K. Nakamoto (eds.), *Introductory Raman Spectroscopy* (Academic Press, Orlando, 1994)
40. T. Jawhari, Micro-raman spectroscopy of the solid state: applications to semiconductors and thin films. Analusis **28**(1), 15–21 (2000)
41. T. Hasegawa, J. Nishijo, J. Umemura, Separation of Raman spectra from fluorescence emission background by principal component analysis. Chem. Phys. Lett. **317**(6), 642–646 (2000)

42. T. Dieing, O. Hollricher, J. Toporski (eds.), *Confocal Raman Microscopy*, vol. 158. Springer Series in Optical Sciences, 2011
43. D.C. Koningsberger, R. Prins, *X-ray Absorption: Principles, Applications, Techniques of EX-AFS, SEXAFS and XANES* (Wiley, New York, 1987)
44. D. Attwood, *Soft X-rays and Extreme Ultraviolet Radiation* (Cambridge University Press, Cambridge, 1999)
45. See http://www.chem.ucalgary.ca/research/groups/faridehj/xas.pdf
46. See http://www.esrf.eu/UsersAndScience/Experiments/CRG/BM25
47. G.R. Castro, Optical design of the general-purpose Spanish X-ray beamline for absorption and diffraction. J. Synchr. Radiat. **5**(3), 657–660 (1998)
48. A. Serrano, O. Rodríguez de la Fuente, V. Collado, J. Rubio-Zuazo, C. Monton, G. Castro, M. García, Simultaneous surface plasmon resonance and x-ray absorption spectroscopy. Rev. Sci. Instr. **83**(8), 083101 (2012)
49. See http://cars9.uchicago.edu/ifeffit/BruceRavel/Horae
50. D.C. Harris, *Análisis químico cuantitativo*. Editorial Reverté, S.A
51. H. Raether, *Surface Plasma Oscillations and their Applications*, vol. 9. Physics of Thin Films, G. Hass, ed. (Academic, New York, 1977)
52. H. Raether, E. Kretschmann, Radiative decay of non radiative surface plasmons excited by light. Zeitschrift für Naturforschung A **23**, 2135–2136 (1968)
53. M. Johnson (ed.), *Photodetection and Measurement: Maximizing Performance in Optical Systems* (McGraw-Hill Professional, New York, 2003)
54. R.B. Bardia, L.P. Viñas, *Circuitos y dispositivos electrónicos: Fundamentos de electrónica* (Edicions UPC, 1999)
55. F. Galvez, C. Monton, A. Serrano, I. Valmianski, J. de la Venta, I.K. Schuller, M.A. Garcia, Effect of photodiode angular response on surface plasmon resonance measurements in the Kretschmann-Raether configuration. Rev. Sci. Instr. **83**(9), 093102 (2012)
56. M.L. Meade (ed.), *Lock-in Amplifiers: Principles and Applications* (P. Peregrinus on behalf of the Institution of Electrical Engineers, London, 1983)
57. J. Polo, T. Mackay, A. Lakhtakia, *Electromagnetic Surface Waves: A Modern Perspective* (Elsevier, New York, 2013)
58. W. Knoll, Interfaces and thin films as seen by bound electromagnetic waves. Annu. Rev. Phys. Chem. **49**, 569–638 (1998)
59. J. Homola, S.S. Yee, G. Gauglitz, Surface plasmon resonance sensors: review. Sens. Actuators B Chem. **54**, 3–15 (1999)
60. C. Miller, A. Sharoni, G. Liu, C. Colesniuc, B. Fruhberger, I. Schuller, Quantitative structural analysis of organic thin films: an x-ray diffraction study. Phys. Rev. B **72**(10), 1–6 (2005)
61. A. Serrano, C. Monton, I. Valmiansk, F. Gálvez, A.L. Cortajarena, J. de la Venta, I.K. Shuller, M.A. García, Study of Co-phthalocyanine films by surface plasmon resonance spectroscopy. J. Appl. Phys. **115**(103106) (2014)
62. See http://www.mpip-mainz.mpg.de/knoll/soft/
63. P.B. Johnson, R.W. Christy, Optical constants of the noble metals. Phys. Rev. B **6**, 4370–4379 (1972)
64. A. Serrano, F. Galvez, O. Rodríguez de la Fuente, M.A. Garcia, X-ray irradiation of soda-lime glasses studied in situ with surface plasmon resonance spectroscopy. J. Appl. Phys. **113**(11), 113104 (2013)

Chapter 3
Extended and Localized Surface Plasmons in Annealed Au Films on Glass Substrates

This chapter presents a study of surface plasmons in Au thin films annealed in air at different temperatures and starting from different initial film thicknesses. The thermal treatment promotes the transition from a continuous film to an ensemble of nanoparticles, which is manifested in the extended or localized character of surface plasmons.

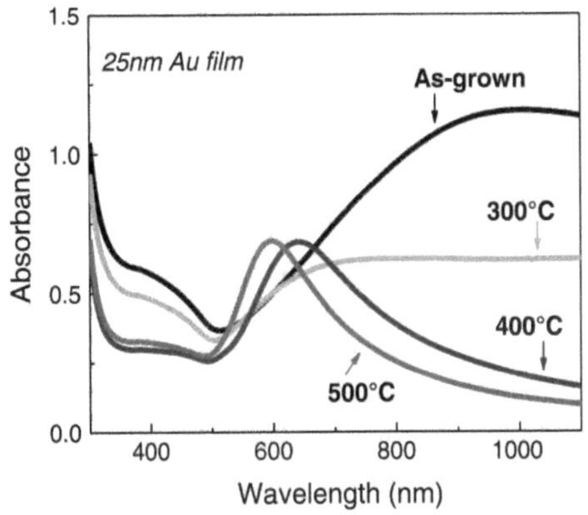

3.1 Introduction

As previously discussed, surface plasmon resonance (SPR) is probably the most outstanding property of noble metal nanostructures. Nanoparticles exhibit localized surface plasmons (LSPs) [1, 2] while thin films hold extended surface plasmons (ESPs) [3–5]. Both kinds of surface plasmons (SPs) have been thoroughly studied, however scarce research has been carried out at the cross-over region where LSPs and ESPs might coexist [6, 7].

A. Serrano Rubio, *Modified Au-Based Nanomaterials Studied by Surface Plasmon Resonance Spectroscopy*, Springer Theses, DOI 10.1007/978-3-319-19402-8_3

As explained in Sect. 2.2.2 of Chap. 2 metallic thin films deposited on substrates with poor adhesion can exhibit substantial morphological modifications when annealing is carried out in air. These modifications include the formation of hillocks, subsequent growing of the holes and island agglomeration leading to a discrete structure [8–12]. The features of both LSPs and ESPs are extremely sensitive to the geometry of nanostructures, exhibiting important changes in the resonant conditions when the size and shape of the nanostructures are modified [1, 13], thus providing a method to tune the optical properties of a given system. Up to now, this method to change the film morphology has been scarcely exploited [12, 14] and has been limited to very thin films to form small nanoparticles with LSPs.

According to this, in this chapter we study changes of both extended and localized surface plasmons resonance of Au thin films annealed in air at different temperatures. The transition from a continuous to a discrete structure is reflected in the SPs, which switch from an extended to a localized character. The features of the SPs are determined by the initial film thickness and the annealing conditions [15].

3.2 Fabrication of Samples

Samples were prepared in two steps. First, Au films were grown onto soda-lime glass substrates by thermal evaporation and second, thermal treatments of Au films were performed. Used substrates were cleaned prior to deposition by subsequent immersion in trichloroethylene, acetone, ethanol and distilled water, and finally dried with N_2 flux. Deposition of Au films was carried out using the commercial Pfeiffer 306 auto coater, in the Departamento de Física de Materiales at the Universidad Complutense of Madrid. Substrates were placed 10 cm away from the tungsten filament and film deposition was performed under a 10^{-6} torr pressure with currents of 34 Amps at a rate of 0.2 nm·s^{-1}. Varying the evaporation time, Au films with thicknesses of 18, 25, 45, 70 and 90 nm were obtained. The thickness of the films was determined through the method explained in Sect. 2.6 of Chap. 2, using the optical absorption of Au films and integrating in the range between 400 and 700 nm.

Thermal treatments of Au films were carried out in air in three steps:

(a) an initial step of 2 h where temperature was gradually increased from 25 °C to the target temperature
(b) a second step of 3 h at constant temperature (300, 400 or 500 °C)
(c) a last cooling step of 2 h,[1] where temperature gradually decreased until the sample reached room temperature

[1]In the cooling step, the oven was programmed to reach the room temperature in 2 h. However, this was not achieved until 8 hours later, approximately.

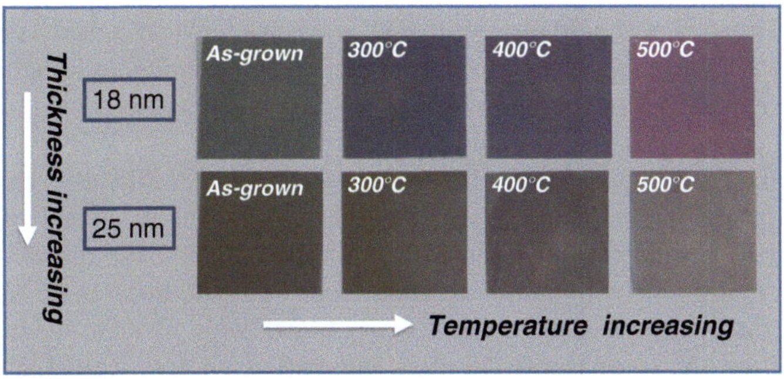

Fig. 3.1 Photographs of samples grown by thermal evaporation with 18 and 25 nm thickness and after a thermal annealing at different temperatures

Some Au samples obtained by this process are shown in Fig. 3.1. Samples with larger thickness than 25 nm are not shown, because they did not exhibit localized surface plasmons and the substrates did not look colored (apart from the golden color in reflection mode).

3.3 Experimental Details

Samples were characterized by atomic force microscopy (AFM) and reflected light optical microscopy (RLOM). Extended surface plasmon resonance (ESPR) was measured by the attenuated total reflection (ATR) method in the Kretschmann-Raether configuration [5], using the setup described in Chap. 2. Optical absorption spectra of samples were recorded with the Shimadzu UV-1603 double beam spectrophotometer ranging between 300 and 1100 nm.

3.4 Morphological Characterization of Samples

The morphological features of the annealed films depend, mainly, on the initial film thickness and annealing conditions such as time, temperature and atmosphere [8]. However, previous studies have shown that an almost stationary situation is achieved for annealing times over 120 min [8–10]. Thus, the time of our annealing processes is assumed to be large enough to reach the almost stationary situation. Moreover, samples were annealed in air accelerating the morphological modifications and favoring the surface diffusion mechanism [10]. Therefore, we focus our study on the effect of initial thickness and annealing temperature [15].

In AFM images, we could observe that all as-grown films presented a continuous structure, while after annealing Au films exhibited a discontinuous structure with a similar phenomenology, although showing different quantitative results, depending on the film thickness and temperature of thermal treatment.

AFM images for the 25 nm Au film are shown in Fig. 3.2. The as-grown sample exhibits a polycrystalline structure with a grain size of about 30 nm and a root mean square (RMS) roughness of 4.6 nm. After annealing at 300 °C, the film exhibits the presence of grains with an almost continuous structure that can be resolved in the image. Grain size distribution results very broad, with grains in the range from 100 to 400 nm. At some points the glass substrate is observed, although the area of uncovered substrate scarcely reaches 10 %[2] of the total specimen. Annealing at 400 °C induces the grain growth with an average size of about 300–400 nm, 80–120 nm height and an uncovered area of around 20 % of the surface. Increasing the annealing temperature up to 500 °C leads to even larger grains with a typical lateral size of 500 nm and a height of 100–120 nm. A larger fraction of uncovered surface, of around 30 %, is seen as well. For this annealing temperature, the uncovered areas percolate and the Au islands become mainly isolated. The Au volume in the annealed samples differs less than 10 % with respect to the initial volume of Au films and it does not show a clear tendency, suggesting that these differences may be attributed to experimental uncertainties. Thus, we may conclude that there is no significant Au removal during the annealing process within our resolution.

The AFM images corresponding to the film with 45 nm thickness are shown in Fig. 3.3. The as-grown film exhibits morphology very similar to that of 25 nm thickness, with a RMS roughness of 1.4 nm and a grain size of about 16 nm. Annealing of Au film at 300 °C leads to the formation of hillocks with hill-to-valley height differences of the order of 20 nm. Due to the larger film thickness, at this temperature there is neither grain separation nor holes nucleation, as evidenced by the AFM image. Annealing at 400 °C promotes the formation and percolation of holes and the growth of Au islands leading to a discrete structure of the film. For this temperature, the average size of the Au islands is of the order of the micron, the uncovered area is roughly the 50 % of the sample surface and most of the islands exhibit elongated shape with an average height of 150 nm. Finally, Au film annealed at 500 °C shows an uncovered area up to 70 % of the sample, the island height increases to an average value of 200 nm, the island size is of the order of the micron and the Au islands become more rounded.

The islands formed from the Au films with the largest initial thicknesses can be observed by optical microscopy. Figure 3.4 shows the optical images obtained by RLOM for the 45 and 90 nm thickness samples annealed at 500 °C. The optical image of Fig. 3.4a and the AFM image of Fig. 3.3g correspond to the same sample.

Films with initial thicknesses of 18, 70, and 90 nm were also studied, although AFM images are not shown. The main results derived from their analysis are shown in Fig. 3.5.

---

[2]This is a minimum value of the uncovered surface due to convolution effects of the AFM tip with a finite radius.

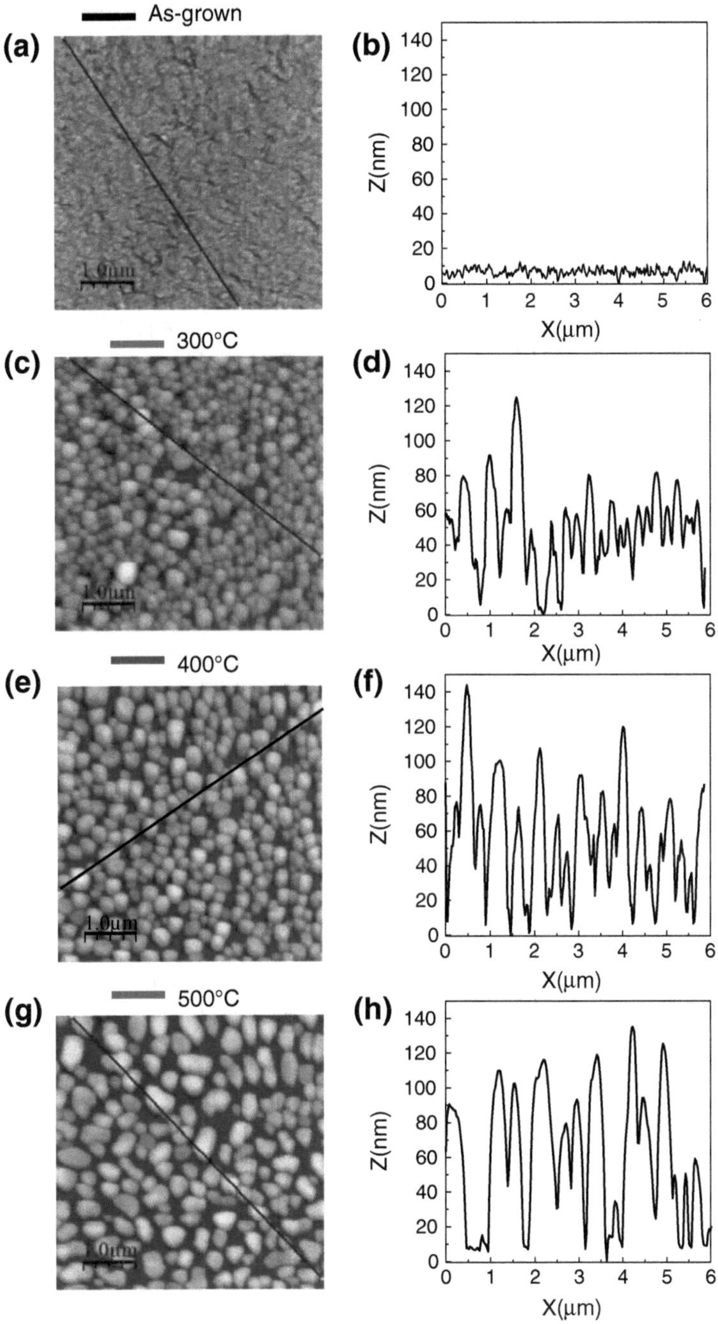

Fig. 3.2 AFM images of Au films with an initial thickness of 25 nm deposited onto soda-lime substrates and annealed at different temperatures in air. *Right panels* present height profiles measured along the lines indicated on AFM images

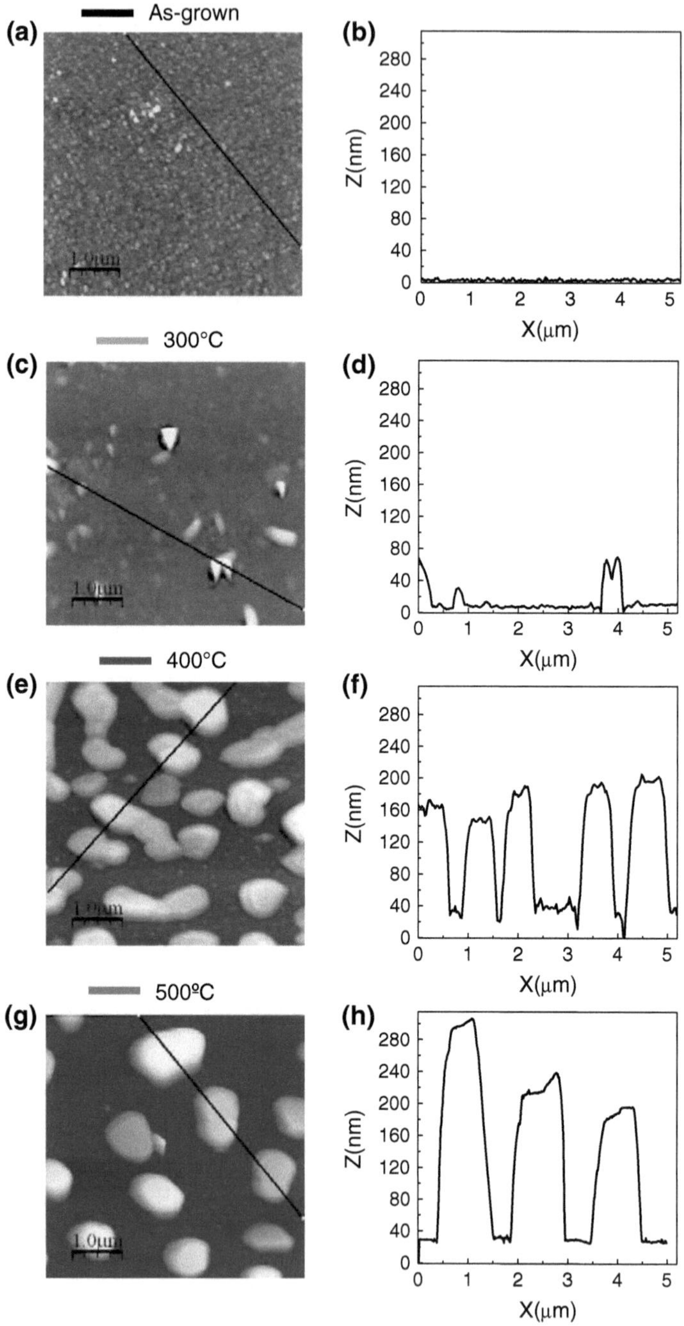

Fig. 3.3 AFM images of Au films with an initial thickness of 45 nm deposited onto soda-lime substrates and annealed at different temperatures in air. *Right panels* present height profiles measured along the lines indicated on AFM images

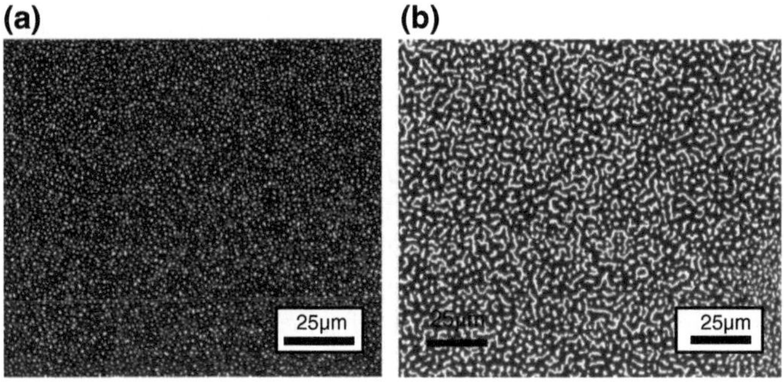

Fig. 3.4 Optical microscopy images of two Au films deposited onto a soda-lime substrate and annealed at 500 °C with an initial thickness of **a** 45 nm and **b** 90 nm. As we work in reflective mode, *bright areas* correspond to gold while *dark areas* correspond to the substrate

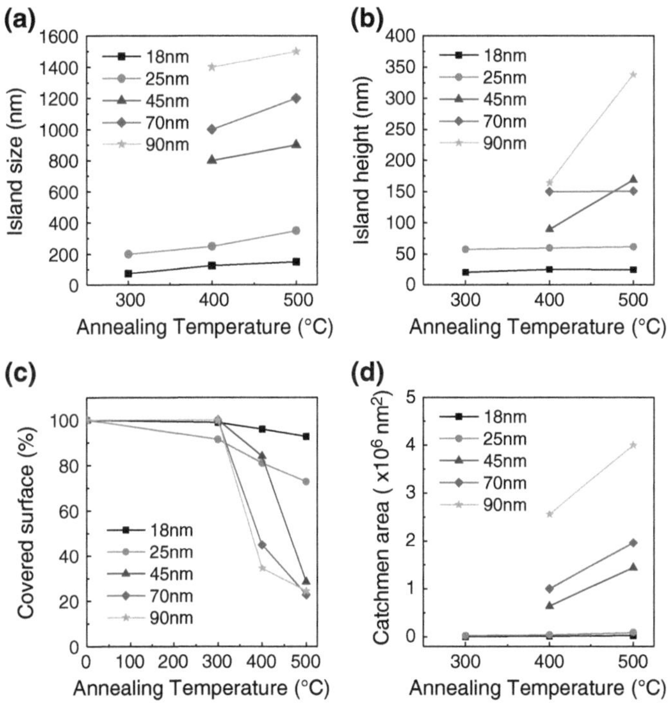

Fig. 3.5 Morphological parameters of the films derived from AFM images as a function of annealing time and initial film thickness: **a** island size, **b** island height, **c** uncovered surface and **d** catchment area

In order to study quantitatively the effect of film thickness and annealing temperature on the Au films morphology, we determined four parameters from the AFM images: island size and height, uncovered surface and catchment area (Fig. 3.5). Catchment area [10] is defined as the area from where the metal atoms migrate to form a single island and is approximately d^2, being d the mean inter-island distance.

As Fig. 3.5 evidences, the morphological features of the samples depend mainly on both annealing temperature and initial film thickness, being the later the most relevant parameter. Modification of morphology in metallic thin films upon annealing has been studied for long time [8, 10–12, 16–18]. Briefly, the main responsible of the process is the reduction of surface and interface energies, which is triggered by the difference in thermal expansion coefficient between the substrate and the metal. As explained in Sect. 2.2.2 of Chap. 2, the inhomogeneous stress distribution at the film/substrate interface promotes the formation of hillocks, but the driving force for dewetting is the minimization of the surface energy of the film and the substrate and that of the film/substrate interface [17, 19]. Higher annealing temperature favors the appearance of holes and their subsequent growth due to the surface diffusion [9]. Finally, holes percolate leading to the formation of metal islands. For annealing temperatures and time large enough, the surface coverage decreases and the Au islands tend to modify their shape becoming more rounded, in order to reduce their surface energy. This behavior is quite expected for metal-on-oxides systems. The most general parameter controlling dewetting is [20]:

$$D = \gamma_s - \gamma_f - \gamma_i \tag{3.1}$$

where γ_s is the substrate surface energy, γ_f is that of the film and γ_i is the interface energy between both materials. If $D < 0$, the continuous film is not stable and dewetting will occur if diffusion is favored. This indeed happens in our system. The surface energy of metals is usually high. On the contrary, surface energy of oxide is usually raether low, as well as that of the interface (adhesion between metals and oxides is poor quite often) [19, 20]. γ_{Au} is around 1.5 J/m^2, γ_{SiO_2} is around 0.3 J/m^2 and γ_{Au/SiO_2} is around 0.6 J/m^2 [21–23].

The formation of the holes becomes easier as Au film thickness decreases, because the amount of material displaced to create a hole is smaller. Specifically, for the 18 and 25 nm films holes were observed after 300 °C annealing, while thicker films showed the presence of holes just upon annealing at 400 °C. For the 18 and 25 nm films, a larger number of holes nucleate and percolate leading to the formation of small islands with limited size, height and catchment area and reducing the covered surface, as Fig. 3.5 shows. However, for films thicker than 25 nm, the fraction of uncovered surface depends mainly on the annealing temperature and it does not depend so significantly on the Au film thickness, provided that the annealing temperature is high enough (see data at 500 °C in Fig. 3.5).

3.5 Optical Characterization of Samples

As we show in this section, the optical measurements present the existence of both extended and localized surface plasmon resonance of samples. The transition from a continuous to a discrete structure is observed in the optical properties switching from an extended to a localized character. The features of the SPs are determined by the initial Au film thickness and the thermal process. In the following sections we present the results obtained.

3.5.1 Extended Surface Plasmon Resonance

Figure 3.6 shows the ESPR spectra for the as-grown and annealed Au films. As explained in Chap. 1, the transmission of the evanescent field of the incident light decreases exponentially across the metal film leading to an electromagnetic field at the metal/dielectric interface. For films thicker than 70–80 nm the electromagnetic field reaching the interface is too weak, while films that are too thin exhibit reduced resonance, due to SPR is damped and there is no chance to excite them effectively.

Comparing the as-grown samples with thicknesses of 25, 45, 70 and 90 nm, we observe the SPR curve is more intense for the 45 nm Au film (the optimum thickness is around 50 nm) than for the rest of samples. For the 25 nm film, the resonance curve is very wide (the electron oscillation is damped) while 70 and 90 nm Au films exhibit a narrower but weaker ESPR curve.

Calculating the ESPR curve for the Au films with the same thickness than the as-grown samples [24], we can observe that the dependence of the measured resonance curve upon film thickness is in agreement with the simulated spectra, shown in Fig. 3.7. The differences in intensity and width between the calculated and experimental spectra can be explained by the surface roughness of the films. This roughness is known to damp to ESPR bands reducing the intensity and widening the absorption band [4].

Samples annealed at 300 °C exhibit wider and weaker ESPR curves than the corresponding to as-grown samples, for all the samples except that of 90 nm. The presence of the ESPR band after annealing at 300 °C is in agreement with the AFM images showing that films exhibited a continuous structure after annealing at this temperature. The formation of hillocks at 300 °C increases the surface roughness leading to damped plasmons and increasing of the width of the resonance. Specifically, for the sample of 90 nm the thermal treatment at 300 °C increases the intensity but the absorption band is wider. The increase in SPR intensity after annealing in the 90 nm Au film could be related to the decrease of film thickness when hillocks are formed after annealing.

Upon annealing at 400 and 500 °C, the ESPR band is practically inappreciable in the spectra, irrespective of the initial film thickness, as expected for films showing a discrete structure with dimensions significantly smaller than light wavelength

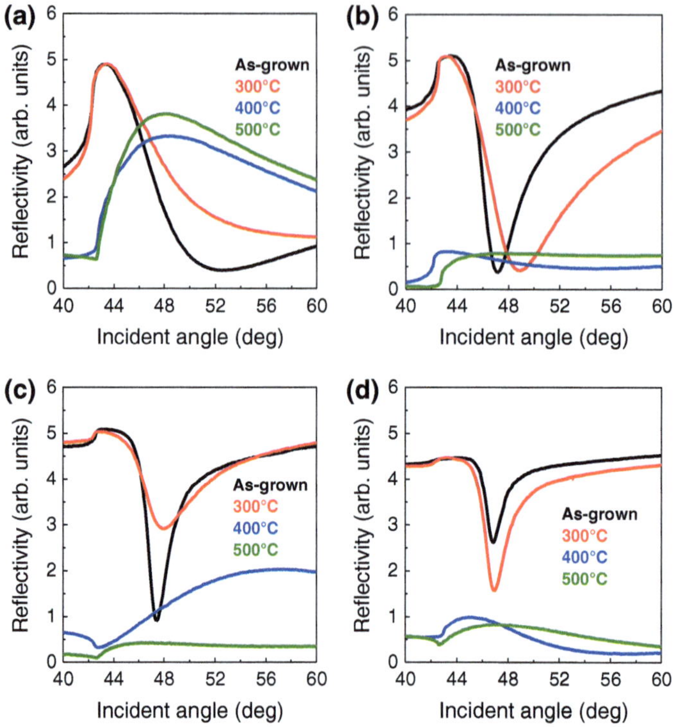

Fig. 3.6 ESPR curves measured in the ATR mode in the Kretschmann-Raether configuration for the Au films with different thickness **a** 25 nm, **b** 45 nm, **c** 70 nm and **d** 90 nm, as-grown and upon annealing in air at different temperatures

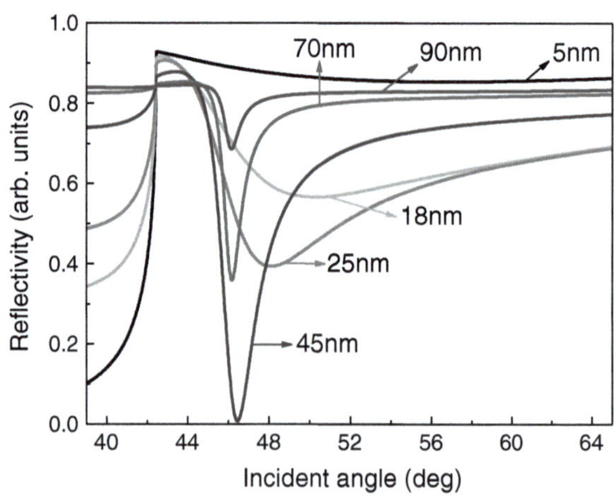

Fig. 3.7 Calculated SPR spectra for Au films ($\varepsilon_1 = -11.74$ and $\varepsilon_2 = 1.25$, according to Johnson and Christy [25]) with different thickness onto soda-lime glass ($n = 1.513 at \lambda = 645\,nm$)

(see Fig. 3.5). SPR measurements in the Kretschmann-Raether configuration can just excite the ESPR in continuous films, but the absence of resonance in these samples does not imply that SPs cannot be excited in the obtained nanostructures by other methods. Actually, patterned films exhibit SPR with features that depend on the patterning geometry [3, 26]. However, the spectra for 45 and 90 nm films annealing at 400 °C still exhibit an ESPR, although the absorption band is very weak and wide (see Fig. 3.6b, d). For these samples, the island size is of 800 nm for 45 nm sample and 1400 nm for 90 nm sample, and the height is of 90 nm for 45 nm sample and 160 nm for 90 nm sample (see Figs. 3.5a, b). With these thicknesses, it is still possible to excite ESPR in the used configuration. Moreover, the heights are an average of the AFM images, but there are islands with smaller heights. The island sizes of 800 nm and 1400 nm are larger than the light wavelength and the large anisotropy of the islands (see Fig. 3.3e) can lead to islands large enough to hold SPR in certain directions with a dispersion relation similar to that of continuous films. Annealing at 500 °C leads to more rounded islands reducing the larger dimension, therefore reducing the absorption due to ESPR.

3.5.2 *Localized Surface Plasmon Resonance*

Figure 3.8 shows the optical absorption spectra of the studied samples. The absorption edge at about 300 nm is due to the soda-lime substrate and cannot be identified with any feature related to Au islands. The shoulder in the range 400–500 nm is due to Au interband transitions, promoting electrons from the 3d band to the Fermi level [27], while the minimum at about 500 nm is characteristic of bulk gold and responsible of the yellowish coloration of this metal [26].

For the 25 nm film annealed at 400 °C, the absorption band characteristic of Au nanoparticles at about 640 nm is clearly observed. This result is in agreement with the AFM images (see Fig. 3.2e), showing that annealing at this temperature leads to a discrete structure with Au nanoparticles. Annealing at 500 °C yields a blue-shift and a narrowing of the absorption band. Main morphological differences between the 25 nm thickness samples annealed at 400 °C and 500 °C are the inter-particle distance (see Figs. 3.2e, g and 3.8a). The short inter-particle distance in the sample annealed at 400 °C favors the dipolar interaction between adjacent particles, which is known to widen and red-shift the localized surface plasmon resonance (LSPR) [1, 13]. Annealing at 500 °C increases the distance between particles, leading to a LSPR band more similar to that expected for isolated Au nanoparticles.

Samples with initial film thickness 45 nm and above do not show LSPR absorption band, regardless of the studied annealing temperature range. Moreover, we analyzed a 30 nm film, which neither showed the absorption band of LSPR upon annealing at same temperature range. Possibly, at higher temperatures LSPR could have been observed but the soda-lime substrate becomes soft at about 700 °C (see Chap. 2). Thus, in these conditions we can establish 30 nm as the upper thickness limit to obtain Au islands exhibiting LSPR using soda-lime substrates. For thicker samples

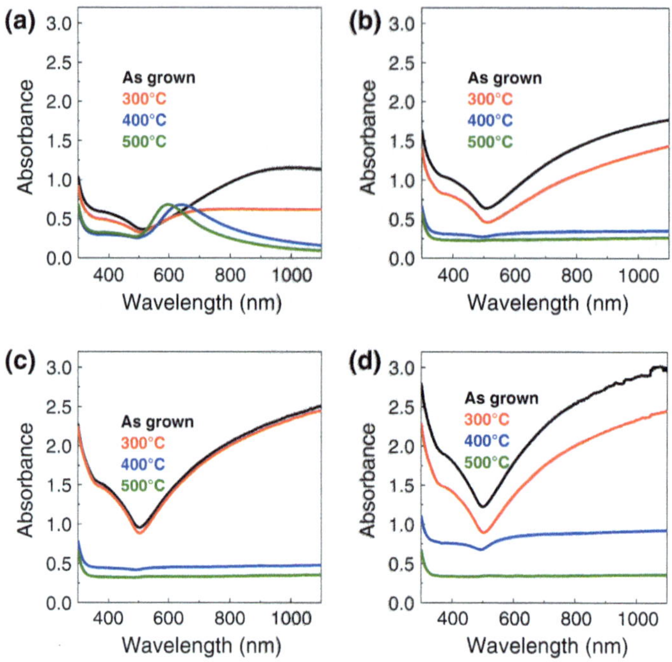

Fig. 3.8 Optical absorption spectra for the Au films with different thickness **a** 25 nm, **b** 45 nm, **c** 70 nm and **d** 90 nm, after annealing at different temperatures

(i.e., 45 nm and above), the main effect of annealing is the decrease of the absorption coefficient in the whole spectral range, which is clearly observed for those films annealed at 400 and 500 °C (blue and green curves in Figs. 3.8b–d). Moreover, a magnified view of these spectra does not reveal any qualitative difference with respect to the corresponding spectra before the annealing. The drastic modification of the absorbance is due to the formation of holes, providing zones where the light beam crosses without interacting with the Au nanostructures. Actually, the evolution of the absorption coefficient is inverse to that of the uncovered surface presented in Fig. 3.9. Thus, we can conclude that the islands exhibit a bulk character from an optical point of view and hence, they do not exhibit the absorption associated with the LSPR.

For the sample with 25 nm initial thickness, we observe a clear transition between ESPR and LSPR. For this thickness, the cross-over region is achieved upon annealing at 300 °C. For this annealing temperature, the spectrum exhibits different features to those expected for pure LSPR and ESPR. The optical absorption spectra of this sample (Fig. 3.8a) does not exhibit the clear maximum at about 600 nm characteristic of LSPR (as those observed for the samples annealed at 400 and 500 °C), but a plateau extended up to the near infrared region (NIR) is observed. Cesario et al. [6] recently demonstrated that the interaction between LSPs and ESPs induces changes in the optical absorption spectra for gold nanoparticles in the proximity of a continuous

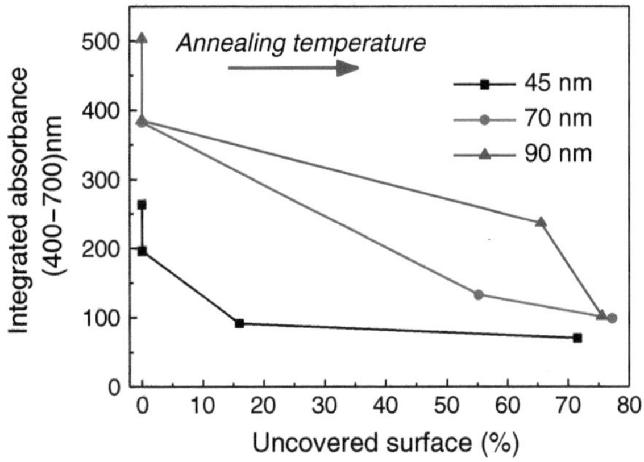

Fig. 3.9 Integrated absorption between (400–700) nm versus the uncovered surface for the films with 45, 70 and 90 nm thickness at different annealing temperatures

silver thin film due to SPR interaction. Reducing the distance between the nanoparticles and the film, the absorption spectra flattens, as we observe here. Concerning the ESPR, the unexpected shift towards larger angles (i.e., larger k value) is observed for the 25 nm sample annealed at 300 °C (Fig. 3.6a), which can also be related to the interaction between LSPR and ESPR. Murray et al. [7] performed dispersion measurements on silver films with holes, ranging from metal percolation to island percolation. They found important changes in the k resonant values in the cross-over percolation region. Therefore, besides the well defined regions of LSPs and ESPs, we find a region where both kinds of excitations may coexist, interacting and providing a new phenomenology.

3.6 Conclusions

Annealing of Au thin films deposited onto soda-lime glass substrates promotes a transition from continuous films to a discrete structure of islands. The features of these islands such as size, shape, height and inter-island distance depend on both the Au film initial thickness and annealing temperature. The structural modifications of the samples lead to substantial changes in their optical properties. The optical response depends mainly on the relative size of Au islands and the light wavelength. Specifically, for films with initial thicknesses up to 30 nm, the Au islands formed after the modification of the film exhibit an optical absorption associated with LSPR. Thicker films lead, upon annealing, to large islands (dimensions larger than visible light wavelength) able to hold ESPR. Therefore, annealing of Au films deposited onto glass substrates provides a method to tune the structural and plasmonic properties of the films in a wide a range and over large areas, not easily achievable by other methods.

References

1. Uwe Kreibig MV (1995) Optical properties of metal clusters, springer series in material science, vol 25. Springer, Berlin
2. E. Hutter, J.H. Fendler, Exploitation of localized surface plasmon resonance. Adv Mater **16**(19), 1685–1706 (2004)
3. M.L. Brongersma, P.G. Kik, *Surface plasmon nanophotonics* (Spring, Berlin, 1988)
4. S.A. Maier, *Plasmonics* (Springer, Berlin, 2006)
5. H. Raether, *Surface plasmons on smooth and rouge surfaces and on gratings* (Springer, Berlin, 1988)
6. J. Cesario, M.U. González, S. Cheylan, W.L. Barnes, S. Enoch, R. Quidant, Coupling localized and extended plasmons to improve the light extraction through metal films. Opt Express **15**(17), 10533–10539 (2007)
7. W.A. Murray, S. Astilean, W.L. Barnes, Transition from localized surface plasmon resonance to extended surface plasmon-polariton as metallic nanoparticles merge to form a periodic hole array. Phys Rev B **69**, 165407 (2004)
8. A.E.B. Presland, G.L. Price, D.L. Trimm, Hillock formation by surface diffusion on thin silver films. Surf Sci **29**(2), 424–434 (1972)
9. S.K. Sharma, J. Spitz, Hillock formation, hole growth and agglomeration in thin silver films. Thin Solid Films **65**(3), 339–350 (1980)
10. M.S.R. Khan, Hillock and island formation during annealing of gold films. Bull Mater Sci **9**, 55–60 (1987)
11. J.A. Thornton, D.W. Hoffman, Stress-related effects in thin films. Thin Solid Films **171**(1), 5–31 (1989)
12. I. Doron-Mor, Z. Barkay, N. Filip-Granit, A. Vaskevich, I. Rubinstein, Ultrathin gold island films on silanized glass. morphology and optical properties. Chem Mater **16**(18), 3476–3483 (2004)
13. M.A. García, J. Llopis, S.E. Paje, A simple model for evaluating the optical absorption spectrum from small au-colloids in sol-gel films. Chem Phys Lett **315**(5–6), 313–320 (1999)
14. S. Pillai, K.R. Catchpole, T. Trupke, M.A. Green, Surface plasmon enhanced silicon solar cells. J App Phys **101**(9), 093105 (2007)
15. A. Serrano, O. Rodríguez de la Fuente, M.A. García, Extended and localized surface plasmons in annealed Au films on glass substrates. J Appl Phys **108**(7), 074303 (2010)
16. S.J. Hwang, J.H. Lee, C.O. Jeong, Y.C. Joo, Effect of film thickness and annealing temperature on hillock distributions in pure Al films. Scripta Mater **56**(1), 17–20 (2007)
17. M.S. Jackson, C.-Y. Li, Stress relaxation and hillock growth in thin films. Acta Metall **30**(11), 1993–2000 (1982)
18. J.V. Coe, J.M. Heer, S. Teerers-Kennedy, H. Tian, K.R. Rodriguez, Extraordinary transmission of metal films with arrays of subwavelength holes. Annu Rev Phys Che **59**, 179–202 (2008)
19. C.V. Thompson, Solid-state dewetting of thin films. Annu Rev Mater Res **42**(1), 399–434 (2012)
20. Rosen MJ (2004) Surfactants and interfacial phenomena. Hoboken, New Jersey: Wiley-Interscience (3rd ed.)
21. H.L. Skriver, N.M. Rosengaard, Surface energy and work function of elemental metals. Phys Rev B **46**(11), 7158–7168 (1992)
22. Brinker CJ, Scherer G W (eds) (1990) Sol-gel science: the physics and chemistry of sol-gel processing. Elsevier
23. M.S. Kennedy, N.R. Moody, D.P. Adamsc, M. Clift, D. Bahra, Environmental influence on interface interactions and adhesion of au/sio$_2$. Mater Sci Eng: A **493**, 299–304 (2008)
24. See http://www.mpip-mainz.mpg.de/knoll/soft/
25. P.B. Johnson, R.W. Christy, Optical constants of the noble metals. Phys Rev B **6**, 4370–4379 (1972)
26. S. Norrman, T. Andersson, C.G. Granqvist, O. Hunderi, Optical properties of discontinuous gold films. Phys Rev B **18**, 674–695 (1978)
27. A. Mooradian, Photoluminescence of metals. Phys Rev Lett **22**, 185–187 (1969)

Chapter 4
Complex Au/FeO$_x$ Nanostructures Obtained from Annealed Bilayers

This chapter presents the fabrication of complex nanoparticles from Au/Fe bilayers, under various configurations and annealing conditions. This new method is applied to obtain successfully Au/α–Fe_2O_3 and Au/α-Fe_2O_3/γ-Fe_2O_3 nanostructures. Their morphological features depend on the layer initial thicknesses and the annealing conditions, which modify the optical absorption band associated with the localized surface plasmons.

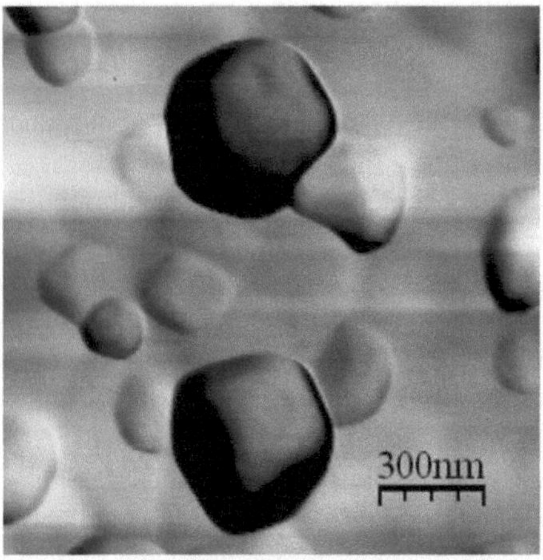

4.1 Introduction

It is well known that nanoparticles (NPs) exhibit different properties than bulk materials with the same chemical composition due to size, surface and proximity effects [1, 2]. Therefore, these nanomaterials offer the possibility to tune their physical properties, such as their optical properties, by controlling size, shape or aggregation states,

A. Serrano Rubio, *Modified Au-Based Nanomaterials Studied by Surface Plasmon Resonance Spectroscopy*, Springer Theses, DOI 10.1007/978-3-319-19402-8_4

yielding many applications in different fields [1, 3, 4]. Complex NPs as core-shell, nano-agglomerates or nano-dimers, exhibit further advantages, since they can combine the properties of their individual components with additional ones arising from their interaction because of the interface and proximity effects [4, 5]. The fabrication of simple NPs (i.e., with a single phase) by chemical or physical routes is relatively well established. It is possible to obtain large amounts of NPs at reasonable cost and with a good control of their size, shape or aggregation [6]. However, fabrication of complex NPs is an open issue not completely solved. Both chemical and physical methods have shown their ability to obtain these complex nanostructures [7–9], but with high cost and low production efficiency. Hence, the fabrication of large amounts of complex NPs at economically viable costs is a milestone in the development of nanotechnology.

As already described in Chap. 2, a well known method to obtain large amounts of NPs dispersed over a substrate consists in the deposition of a metallic thin film and a subsequent annealing [10, 11]. The difference in thermal expansion coefficient between the metallic film and the substrate induces mechanical stresses during the annealing process. As a result, the metallic film undergoes morphological modifications in order to relax this stress: formation of hillocks (decreasing the surface energy and that of the film-substrate interface) [12, 13], hole nucleation with increasing temperature and subsequently agglomeration leading to the formation of islands or nanostructures [10, 14, 15]. Until now, this method has been successfully used to obtain noble metal nanoparticles [15, 16]. However, few works have explored the possibility to use it with other non-noble metals [17–19]. Moreover, to our best knowledge, this method has never been applied to obtain complex NPs with more than one phase.

According to this, in this chapter, we address the fabrication and characterization of complex NPs over silica substrates by deposition of bilayers or thin films and subsequent annealing. We focus on the system Au/FeO_x because of the outstanding properties of each of the constituent elements. On the one hand, as indicated in Chap. 1, Au NPs can exhibit intense optical activity associated with the surface plasmons (SPs), chemical stability, high biocompatibility and easy functionalization [1, 20]. On the other hand, iron oxide (FeO_x) can present ferrimagnetism, low toxicity and intense photocatalytic activity [21]. Besides this, the combination of those components may exhibit additional features as enhanced photocatalytic activity for CO oxidation [22] or coupling between magnetic and optical properties of the elements [23] due to surface or proximity effects. Finally, the features related to SPs are evaluated as a function of the annealing conditions and the thickness of initial films.

4.2 Experimental Details

Au and Fe films were deposited onto silica substrates by thermal evaporation using the home-made evaporation chamber mounted in the Departamento de Electrocerámica at the Instituto de Cerámica y Vidrio (CSIC) of Madrid (see Chap. 2). Prior to the deposition, substrates were cleaned with soap and water, and dried with dry air flux. Substrates were placed about 25 cm away from the "V"-shaped tungsten filaments, where Au and Fe wires were initially melted to form a ball. One tungsten filament for each material was used with its own evaporation source. The deposition was performed under a 10^{-6} torr pressure and the deposition rate was about 0.02 nm/s, controlled with a Q-microbalance. Afterwards, metallic films were annealed at different conditions (time, temperature, atmosphere), in order to obtain Au/FeO_x nanostructures. Thermal treatments consisted of an initial step from room temperature (RT) to the target temperature, a second step where the target temperature (which depends on each material) was constant for a certain time and a last cooling step where the temperature decreased until RT.

Samples were characterized by atomic force microscopy (AFM), scanning electron microscopy (SEM), reflected light optical microscopy (RLOM) and confocal Raman microscopy. Optical absorption spectra of the samples were recorded using the V-670 UV-Visible double beam spectrophotometer.

4.3 Au/FeO_x Nanostructures Obtained by Annealing in Air

4.3.1 Fabrication of Samples Prepared in Air

The temperature required for breaking continuous thin films into nanostructures depends on the particular metal, annealing process and substrates used [14]. The difference in thermal expansion coefficient between the film and the substrate induces thermal stresses at the interface that are relieved breaking the film into NPs. For Au films onto glass substrates, the large difference between linear expansion coefficient of metallic film and the glass substrate [10, 15, 24, 25] (see Chap. 2) can prompt the appearance of significant stresses and the formation of NPs at temperatures as low as 200 °C. In the case of Fe films, the formation of NPs by this method has not been almost studied [18]. Fe has a thermal expansion coefficient similar to that of Au [24], but becomes oxidized in few minutes upon annealing at 150 °C for 1 h [26], a temperature too low to promote the formation of NPs. Thus, prior to the formation of metallic Fe NPs, we have an FeO_x film. The thermal expansion coefficient of iron oxides is significantly smaller than that of the metallic Fe (i.e., 8.10^{-6} K^{-1} for α-Fe_2O_3) [24]. Consequently, higher temperatures are required to achieve the strains needed for the formation of the NPs from an Fe film deposited on a glass substrate, in comparison with those for the fabrication of Au NPs. Such stresses are only obtained above 900–1200 °C, in which an Fe film has completely transformed into iron oxide NPs, where the oxidation state depends on the atmosphere (O_2 pressures) during the thermal treatment [27]. At these temperatures, the Au is melted, making difficult the fabrication of Au/FeO_x NPs. Due to the high temperatures required to obtain iron oxide NPs, for these experiments, we chose silica substrates because they present a higher softening point (1600 °C) than the soda-lime substrates (700 °C) used in the previous chapter (see Sect. 2.2.2 of Chap. 2).

Annealing in high vacuum or reducing atmospheres in order to prevent iron oxidation is not a chance. The mass transport mechanism for the formation of NPs is mainly the surface diffusion [10]. The surface diffusion coefficient of metals is increased several orders of magnitude by the oxygen adsorbed on the metal surface [28, 29]. Therefore, NPs cannot be obtained following this method without oxidising the Fe film.

The large difference in the temperatures required to obtain Au (from 200 °C) and FeO_x NPs (from 900 °C) determines the strategy to fabricate the hybrid nanostructures. Thus, two different approaches were followed to obtain complex Au/FeO_x nanostructures by thermal treatments in air:

- *First approach*: initially, Fe films were grown on silica substrates and annealed at 1050 °C for 4 h in air to obtain the FeO_x NPs. Afterwards, Au films were deposited over the FeO_x nanostructured film and a second thermal treatment at 500 °C for 3 h in air was carried out. As the temperature of this second annealing is significantly smaller than the initial one, the FeO_x NPs do not suffer appreciable modifications during the annealing process. Within this approach, the heating and cooling steps of the thermal treatments were carried out in 2 h until or from the corresponding target temperature. Thermal treatments were similar to those used to obtain Au NPs in Chap. 3. In this way, we prepared samples where the Fe film thickness was varied: 5, 10 and 20 nm, while the Au thickness, 10 nm, and the annealing conditions remained fixed.
- *Second approach*: Fe/Au bilayers, where the Au films are in contact with the silica substrates, were grown and a subsequent single annealing at higher temperature for a short time was carried out, in order to obtain Au/FeO_x nanostructures in just one annealing. In this case, the annealing temperature was varied from 1050 to 1200 °C, in order to obtain more isolated nanostructures than in the first approach. At 1200 °C, it is possible to obtain FeO_x NPs. However, this temperature is too high to fabricate Au NPs (the melting point of Au is 1064 °C, see Chap. 2) since the Au would not be in solid state, which it is necessary to obtain NPs from films [13]. For this reason, we shorten the annealing time, which was of the order of minutes, instead of hours and during the cooling step, the temperature was quickly decreased: in 6 min from the target temperature of 1200 °C to 600 °C and from 600 °C to RT in 40 min. The heating step was similar to the previous approach: in 2 h, temperature was gradually increased from RT to the target temperature. Besides, Au films were deposited before the Fe films, fabricating Fe/Au bilayers instead of Au/Fe bilayers in order to protect the Au films. With this configuration, the amount of oxygen that reaches the Au is much smaller than when the Au layer is completely uncovered. Thus, within this second approach, the annealing parameters were chosen since they supposed a compromise to obtain both kinds of NPs. In this way, we prepared samples varying the Fe film thickness: 2, 5, 10 and 20 nm and the annealing time: 3 or 30 min, while the rest of the parameters (i.e., annealing temperature at 1200 °C and Au film thickness of 10 nm) remained fixed.

As indicated above, few works have explored the possibility of obtaining iron oxide nanostructures from Fe films [17–19]. Thus, prior the fabrication of Au/FeO_x

NPs, we prepared FeO_x NPs starting from Fe thin films annealed at 1050 °C for 4 h in air. These annealing conditions were chosen because Aggarwal et al. reported the fabrication of α-Fe_2O_3 NPs from Fe films annealed at 1050 °C [18, 30]. For the obtention of FeO_x NPs, Fe thin films varying the thickness (5, 10, 15 and 20 nm) were grown. In both the heating and cooling steps of the thermal treatments, temperature was gradually increased and decreased in 2 h.

The area of the substrates that were coated with the metallic films was of 1 × 1 cm^2. Table 4.1 shows a short description of the FeO_x and Au/FeO_x NPs obtained in air.

Table 4.1 FeO_x and Au/FeO_x NPs fabricated from Au and Fe films with different thickness and under different annealing conditions in air

Sample	Description
FeO_x NPs	
Fe 5 1050 °C 4 h	5 nm Fe film annealed at 1050 °C for 4 h
Fe 10 1050 °C 4 h	10 nm Fe film annealed at 1050 °C for 4 h
Fe 15 1050 °C 4 h	15 nm Fe film annealed at 1050 °C for 4 h
Fe 20 1050 °C 4 h	20 nm Fe film annealed at 1050 °C for 4 h
Au/FeO_x NPs—*First approach*	
Fe 5 1050 °C Au 10 500 °C	5 nm Fe film annealed at 1050 °C for 4 h plus 10 nm Au film annealed at 500 °C for 3 h
Fe 10 1050 °C Au 10 500 °C	10 nm Fe film annealed at 1050 °C for 4 h plus 10 nm Au film annealed at 500 °C for 3 h
Fe 20 1050 °C Au 10 500 °C	20 nm Fe film annealed at 1050 °C for 4 h plus 10 nm Au film annealed at 500 °C for 3 h
Au/FeO_x NPs—*Second approach*	
Au 10 Fe 2 1200 °C 3 min	2 nm Fe/10 nm Au bilayer annealed at 1200 °C for 3 min
Au 10 Fe 5 1200 °C 3 min	5 nm Fe/10 nm Au bilayer annealed at 1200 °C for 3 min
Au 10 Fe 10 1200 °C 3 min	10 nm Fe/10 nm Au bilayer annealed at 1200 °C for 3 min
Au 10 Fe 20 1200 °C 3 min	20 nm Fe/10 nm Au bilayer annealed at 1200 °C for 3 min
Au 10 Fe 2 1200 °C 30 min	2 nm Fe/10 nm Au bilayer annealed at 1200 °C for 30 min
Au 10 Fe 5 1200 °C 30 min	5 nm Fe/10 nm Au bilayer annealed at 1200 °C for 30 min
Au 10 Fe 10 1200 °C 30 min	10 nm Fe/10 nm Au bilayer annealed at 1200 °C for 30 min
Au 10 Fe 20 1200 °C 30 min	20 nm Fe/10 nm Au bilayer annealed at 1200 °C for 30 min

4.3.2 Morphological Characterization of Samples Prepared in Air

- **FeO_x nanostructures**

Figure 4.1 shows the AFM images corresponding to the sample of FeO_x NPs obtained by deposition of a 15 nm Fe film and a consequent annealing at 1050 °C for 4 h in air, sample Fe 15 1050 °C 4h. The annealing process induces quite rounded NPs, which are distributed homogeneously throughout the whole substrate surface. Some areas of the silica substrate can be distinguished. The NP size ranges between 100 and 500 nm, and the NP height is of 50 nm. For the rest of fabricated FeO_x NPs using the same annealing conditions and varying the initial Fe thickness between 5 and 20 nm, the morphology is qualitatively quite similar to that of sample Fe 15 1050 °C 4h, but NPs are more agglomerated and their sizes and heights increase as

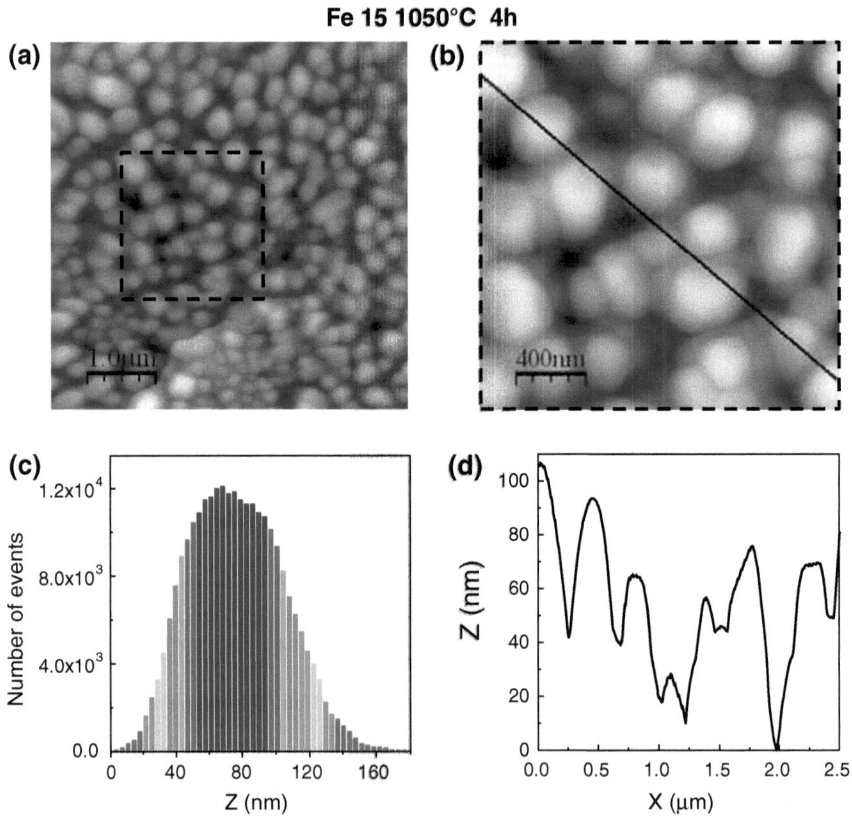

Fig. 4.1 **a** and **b** AFM images of 15 nm thickness Fe film deposited onto a silica substrate and annealed at 1050 °C for 4 h in air. **c** Height histogram taken from the AFM image **a** and **d** height profile measured along the line indicated on AFM image

initial Fe thickness increases. In general, structural features such as the size, shape or aggregation state of FeO_x NPs could be modified varying the annealing conditions besides the initial Fe film thickness.

After obtaining FeO_x NPs from Fe thin films annealed in air, we followed the two approaches described above to obtain complex Au/FeO_x nanostructures by thermal treatments in air.

- **First approach: Fe film + first annealing + Au film + second annealing**

Figure 4.2 shows SEM micrographs of two samples prepared following this approach, samples Fe 5 1050 °C Au 10 500 °C and Fe 10 1050 °C Au 10 500 °C (see Table 4.1). These images show the large homogeneity of NPs grown over the silica substrate surface, being possible to distinguish two nanostructured films: a bottom layer consisting of gray color nanostructures and a top film of white color NPs. Since the intensity of the backscattered electron signal can be related to the atomic number of the specimen, it is possible to identify the white NPs as those of Au and the gray NPs as FeO_x ones. For both samples, we can observe that the top layer (Au NPs) presents similar sizes and shapes. Moreover, sample Fe 10 1050 °C Au 10 500 °C shows Au NPs more separated than sample Fe 5 1050 °C Au 10 500 °C.

Figure 4.3 shows an AFM image for the sample Fe 10 1050 °C Au 10 500 °C, where it is possible to differentiate a bottom layer of NPs with a size in the range 50–300 and 10–40 nm height and a second type of larger NPs on top of the previous ones with average size about 500 nm and height of the order of 100–130 nm. Figure 4.4a, c shows a region of FeO_x NPs for samples Fe 10 1050 °C Au 10 500 °C and Fe 20 1050 °C Au 10 500 °C respectively, where it is possible to differentiate NPs with an average size around 200 nm and 10–40 nm height for Fe 10 1050 °C Au 10 500 °C (see Fig. 4.4b) and NPs with an average size around 300 and 20–80 nm height for Fe 20 1050 °C Au 10 500 °C (see Fig. 4.4d). AFM image of sample Fe 5 1050 °C Au 10 500 °C is not shown because it could not be obtained due to tip effects. However, for this sample, we observed that the FeO_x film contains smaller NPs than samples

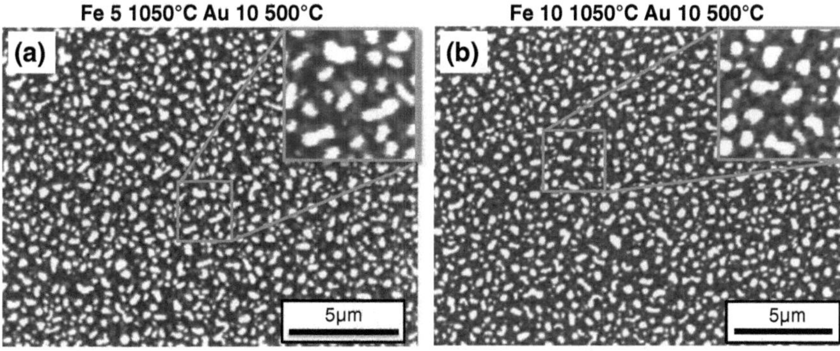

Fig. 4.2 SEM images for an Fe film with **a** 5 nm and **b** 10 nm thickness annealed at 1050 °C for 4 h plus a 10 nm Au film annealed at 500 °C for 3 h in air

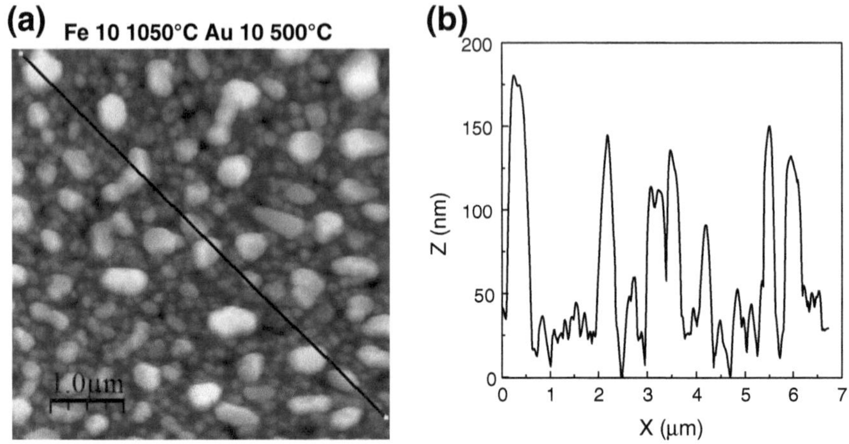

Fig. 4.3 **a** AFM image for a 10 nm Fe film annealed at 1050 °C for 4 h plus a 10 nm Au film annealed at 500 °C for 3 h in air and **b** height profile measured along the line indicated on AFM image

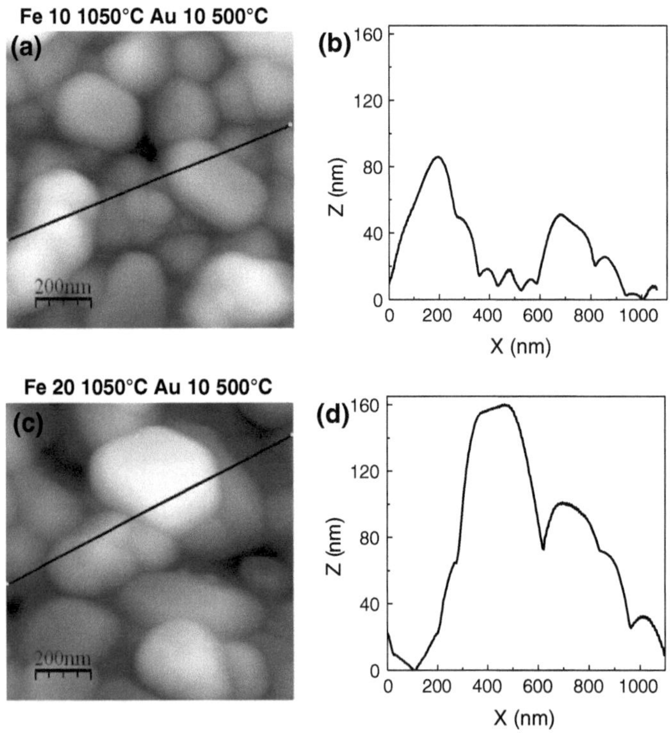

Fig. 4.4 AFM images for an Fe film with **a** 10 nm and **c** 20 nm thickness annealed at 1050 °C for 4 h plus a 10 nm Au film annealed at 500 °C for 3 h in air. (*Right*) height profiles measured along the line indicated on AFM images

grown from the 10 and 20 nm Fe films (see Fig. 4.4). Hence, samples fabricated with this approach show iron oxide NPs smaller and more separated for the thinnest initial Fe films.

It should be noted that the Au NPs obtained on the FeO_x NPs layer result larger than those obtained for a Au film with the same thickness deposited directly on the glass substrate (see Chap. 3). The discontinuous profile of the iron oxide nanostructured film induces a more heterogeneous stress distribution than in a flat glass substrate. This favors the nucleation of holes, leading to larger and more separated Au NPs in order to minimize the interface energy between the FeO_x NPs and the Au film (although the difference between the expansion coefficients between the Au film and the FeO_x nanostructured film is smaller than that between the Au film and the silica substrate). Hence, parameters as the size, shape and aggregation state of FeO_x NPs also determine the morphological features of the Au NPs, modifying their physical properties.

Therefore, with this approach, modifying the thickness of initial Fe films, the morphology of both kinds of NPs and consequently their individual properties can be tuned independently. Other parameters different from initial Fe thickness could be changed.

- **Second approach: Fe/Au bilayer + annealing**

The coupling effects between the Au and FeO_x NPs can be enhanced upon the formation of dimers with both components. The formation of these dimers is quite difficult with the first approach, in which each kind of NPs is processed in separated steps and at fairly different temperatures. In order to promote the formation of these dimers, a second approach was used, which is described above.

Figure 4.5 shows the SEM micrographs of four samples prepared by this second approach, consisting of Fe/Au bilayers annealed at 1200 °C for 3 min in air, fixing the Au film thickness to 10 nm and varying the Fe film thickness: 2, 5, 10 and 20 nm (see Table 4.1). These images confirm that the NPs exhibit a homogeneous size distribution over the substrate surface. Moreover, the NP morphology varies depending on the initial Fe film thickness: the larger the Fe film thickness, the larger the complex nanostructures. In all samples, the bare substrate can be observed. Besides this, we can distinguish NPs with different color: white and gray NPs, which correspond to Au nanostructures and FeO_x ones, respectively.

AFM images for complex nanostructures fabricated with this approach for 3 and 30 min are illustrated in Figs. 4.6 and 4.7, respectively. For both annealing times, the morphological features such as the size, shape or aggregation state of hybrid NPs vary largely when changing the Fe film thickness. Two kinds of NPs can be distinguished: one type of nanostructures is higher than the other one. Phase AFM images (Figs. 4.8 and 4.9) were analyzed and they evidenced the formation of complex dimers. These images were taken scanning on sample surfaces slowly, in order to avoid topographic effects in the phase signal. A detail of AFM images illustrated in Figs. 4.6 and 4.7 (see squares on AFM images) allows intuiting both species by phase contrast. The NPs exhibit a dual distribution of phase contrast than can be associated with two

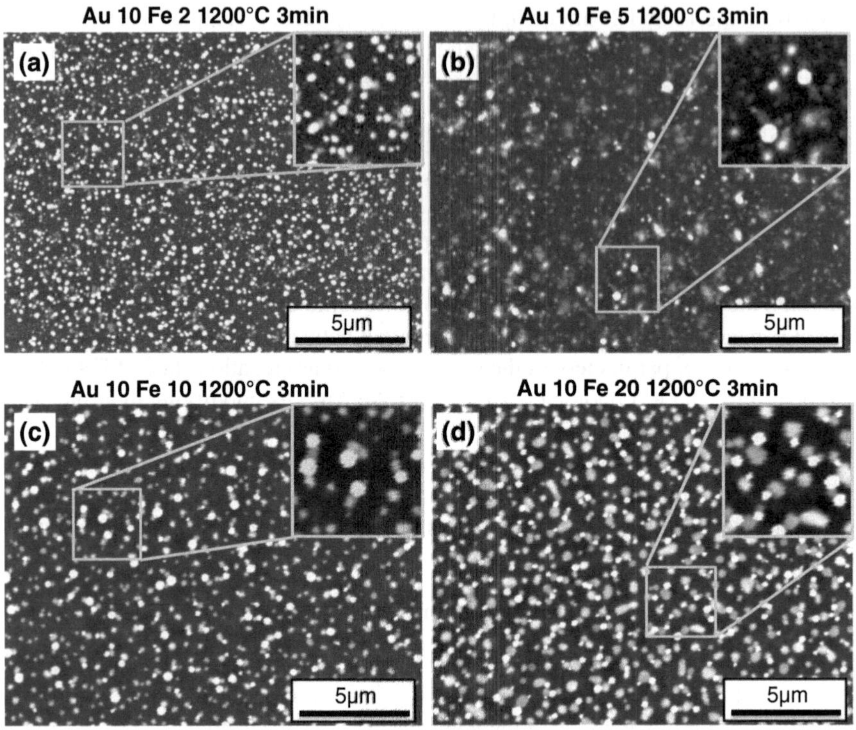

Fig. 4.5 SEM images for samples obtained from Fe/Au bilayers annealed at 1200 °C for 3 min in air, with a 10 nm Au thickness and varying the Fe film thickness: **a** 2 nm, **b** 5 nm, **c** 10 nm and **d** 20 nm

chemical phases. Due to the morphological features of nanostructures with smaller phase contrast (similar for all samples fabricated with this second approach), these may be assigned to Au NPs while those with higher phase contrast may be identified as FeO$_x$ NPs.

As shown Fig. 4.6a, complex NPs in Au 10 Fe 2 1200 °C 3 min present an average size between 50 and 300 nm and a height in the range from 10 to 60 nm, showing a large percentage of bare substrate. Increasing the Fe film thickness from 2 to 5 nm, sample Au 10 Fe 5 1200 °C 3 min (Fig. 4.6b), the NPs present a completely different morphology. In this case, the highest NPs with a size between 200 and 600 nm and a height of 100–150 nm are over a nanostructured film, which shows nanoparticles with an average size around 100–200 nm and a height of the order of 20 nm. In the case of the sample Au 10 Fe 10 1200 °C 3 min (Fig. 4.6c), NPs show a size ranging from 100 to 400 nm and a height from 10 to 140 nm. For the sample with a 20 nm Fe thickness, sample Au 10 Fe 20 1200 °C 3 min (Fig. 4.6d), NPs present an average size in the range from 100 to 500 nm and a height between 20 and 150 nm. For this batch, samples with an initial Fe thickness of 5, 10 and 20 nm show the substrate

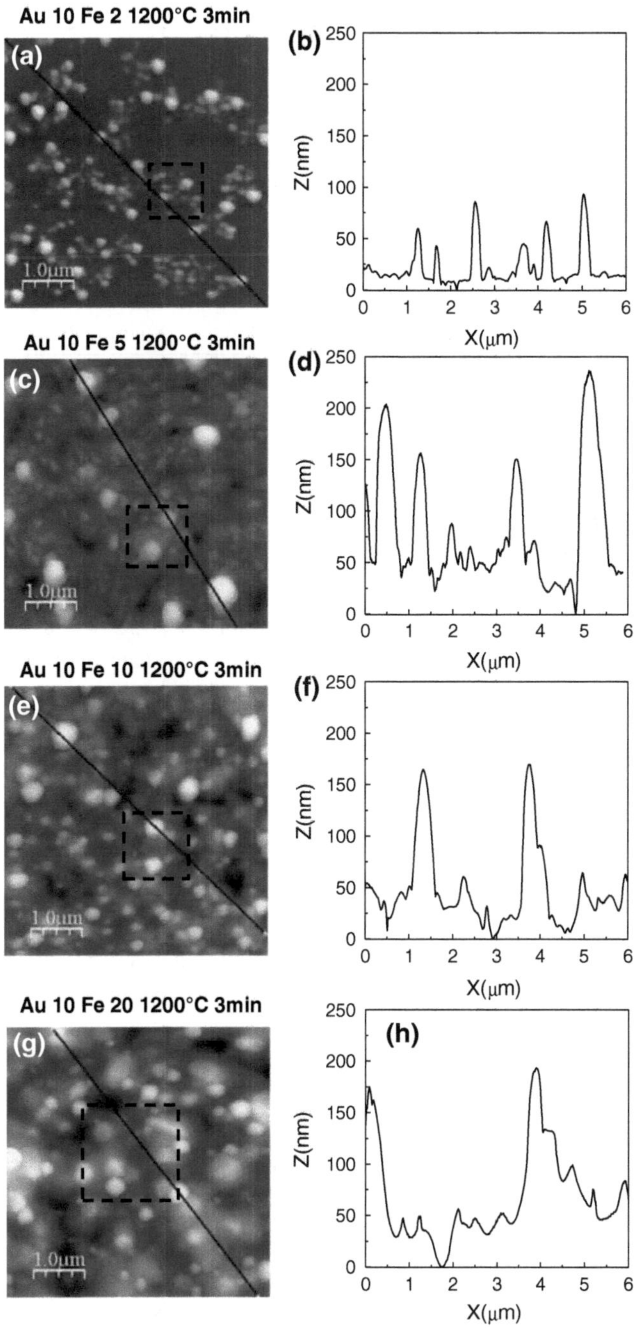

Fig. 4.6 Topography AFM images for samples obtained from Fe/Au bilayers annealed at 1200 °C for 3 min in air, with a 10 nm Au thickness and varying the Fe film thickness: **a** 2 nm, **c** 5 nm, **e** 10 nm and **g** 20 nm. (*Right*) height profiles measured along the lines indicated on AFM images

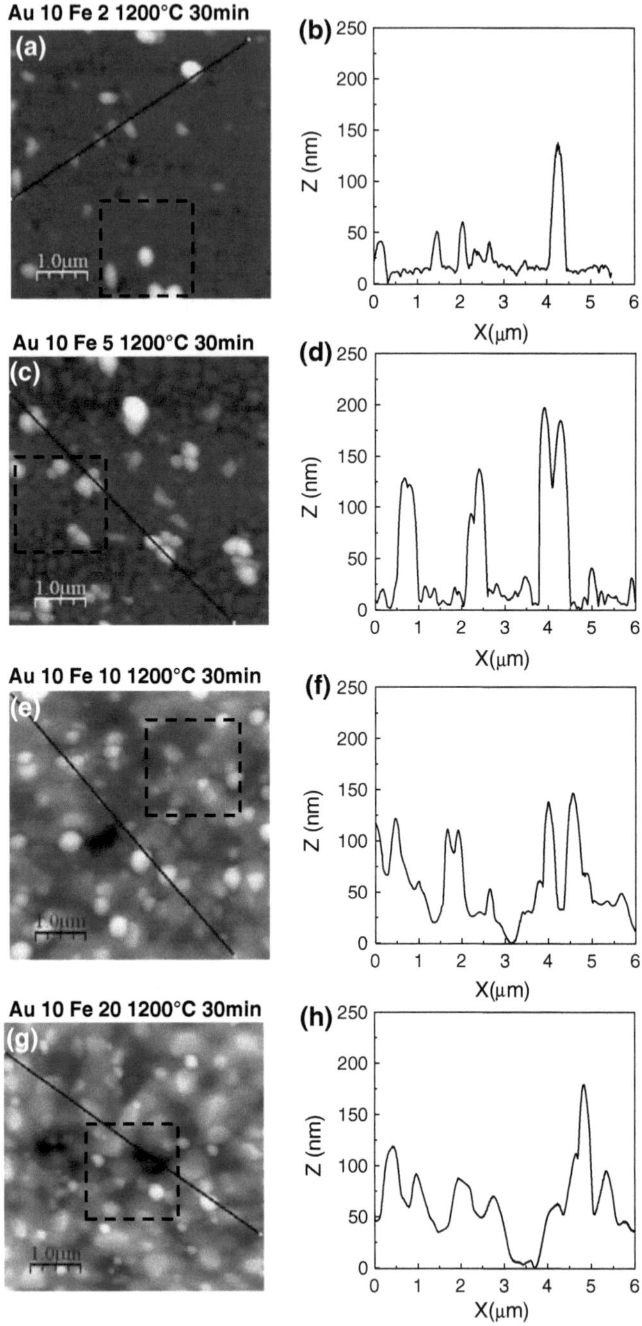

Fig. 4.7 Topography AFM images for samples obtained from Fe/Au bilayers annealed at 1200 °C for 30 min in air, with a 10 nm Au thickness and varying the Fe film thickness: **a** 2 nm, **c** 5 nm, **e** 10 nm and **g** 20 nm. No significant changes are observed with respect to Fig. 4.6. (*Right*) height profiles measured along the lines indicated on AFM images

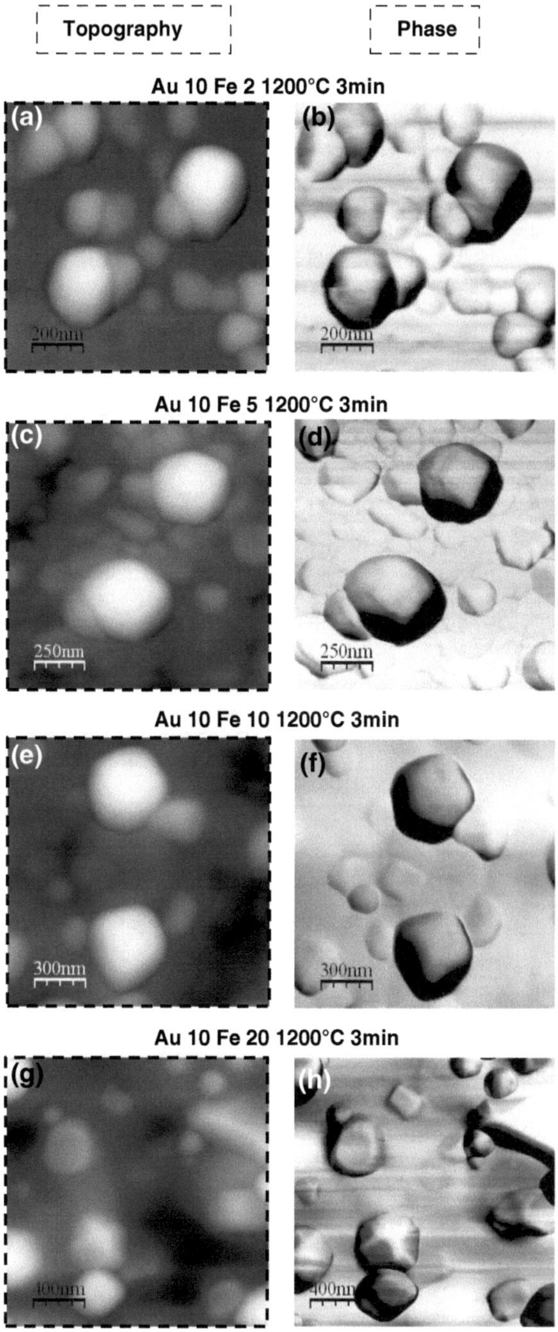

Fig. 4.8 (*Left*) a detail of topography AFM images shown in Fig. 4.6 and (*right*) phase AFM images for samples obtained from Fe/Au bilayers annealed at 1200 °C for 3 min in air, with a 10 nm Au thickness and varying the Fe film thickness: 2, 5, 10 and 20 nm

Fig. 4.9 (*Left*) a detail of topography AFM images shown in Fig. 4.7 and (*right*) phase AFM images for samples obtained from Fe/Au bilayers annealed at 1200 °C for 30 min in air, with a 10 nm Au thickness and varying the Fe film thickness: 2, 5, 10 and 20 nm

surface more covered by NPs, around 5–10 % is uncovered, than for the sample obtained from a 2 nm Fe film, which shows an uncovered area of the order of 70 % (see Fig. 4.6).

Increasing the annealing time, from 3 to 30 min, does not alter significantly the NP morphology (compare Figs. 4.6 and 4.7). For Au 10 Fe 2 1200 °C 30 min (see Fig. 4.7a), complex NPs present an average size in the range from 50 to 400 nm and a height of a few nanometers to 80 nm. For sample Au 10 Fe 5 1200 °C 30 min (Fig. 4.7b), two types of nanostructures can be distinguished: the highest NPs with a height of 100–150 nm and an average size from 200 to 500 nm and the other ones with a height around 20 nm and an average size of 100–200 nm. For sample obtained from a 10 nm Fe film (Au 10 Fe 10 1200 °C 30 min, Fig. 4.7c), the NPs show an average size in the range from 150 to 400 nm and a height between 20 and 100 nm. For the case of the sample Au 10 Fe 20 1200 °C 30 min (see Fig. 4.7d), NPs present an average size between 150 and 500 nm and a height ranging from 20 to 120 nm (see roughness profiles in Fig. 4.7). For this second batch, samples show around 5–15 % of the substrate surface uncovered by NPs. Specifically, complex NPs obtained from Fe/Au bilayers with 2 and 5 nm Fe thicknesses annealed for 30 min present different aggregation states and shapes to those samples annealed for 3 min. NPs obtained from Fe/Au bilayers with 10 and 20 nm Fe thicknesses show similar morphology for both annealing times.

This second method does not allow a complete control of the morphology and structure of both species as the first approach, however it is possible to tune them at somewhat extended in just one thermal treatment. Moreover, this method can favor the interaction between the complex NPs since are shown as dimers.

4.3.3 Confocal Raman Characterization of Samples Prepared in Air

As described in Chap. 2, using Raman spectroscopy it is possible to distinguish different materials and differentiate species in a same sample. A common problem in the Raman measurements is the fluorescence phenomenon of metals such as the Au. In the literature, we can find many works to avoid the fluorescence or to separate the Raman signal of a material from fluorescence emission [31, 32]. However, the fluorescence can also be used to differentiate materials that present this phenomenon from other materials that do not show it. Taking this into account, a study by confocal Raman microscopy of samples was performed mapping areas both on the plane (XY plane) and in depth (XZ plane), in order to determine the phase of iron oxide NPs and trying to distinguish between the different components, separating the fluorescence signal for Au NPs and the signal of FeO_x NPs by their characteristic spectrum. Both FeO_x and complex Au/FeO_x nanostructures prepared by annealing in air were studied.

- **FeO$_x$ nanostructures**

Figure 4.10 shows the Raman spectra and images for FeO$_x$ NPs obtained from a 20 nm Fe film annealed at 1050 °C for 4 h in air, sample Fe 20 1050 °C 4h. In the whole sample, we only identified a type of Raman spectrum. In-plane (10 × 10 μm^2) and in-depth (10 × 2.5 μm^2) Raman intensity images were obtained mapping the FeO$_x$ Raman spectra taken each 300 nm, integrating in the spectral range from 0 to 3600 cm^{-1}. These mapped regions are marked with a purple square (in-plane image, Fig. 4.10b) and a blue line (in-depth image, Fig. 4.10c) on the optical micrograph of the sample (see Fig. 4.10a). In the in-plane Raman image, we can observe the homogeneous dispersion of the FeO$_x$ NPs on the silica surface. Moreover, in the in-depth Raman image, it is possible to differentiate clearly between the NPs and the substrate and estimate the sample thickness, around 600 nm. The single Raman spectra, which represent a pixel in the Raman image, were measured each 300 nm. Besides, the optical resolution of the confocal microscope in the vertical direction is of 500 nm. This implies that the thickness value is larger than the real one and the

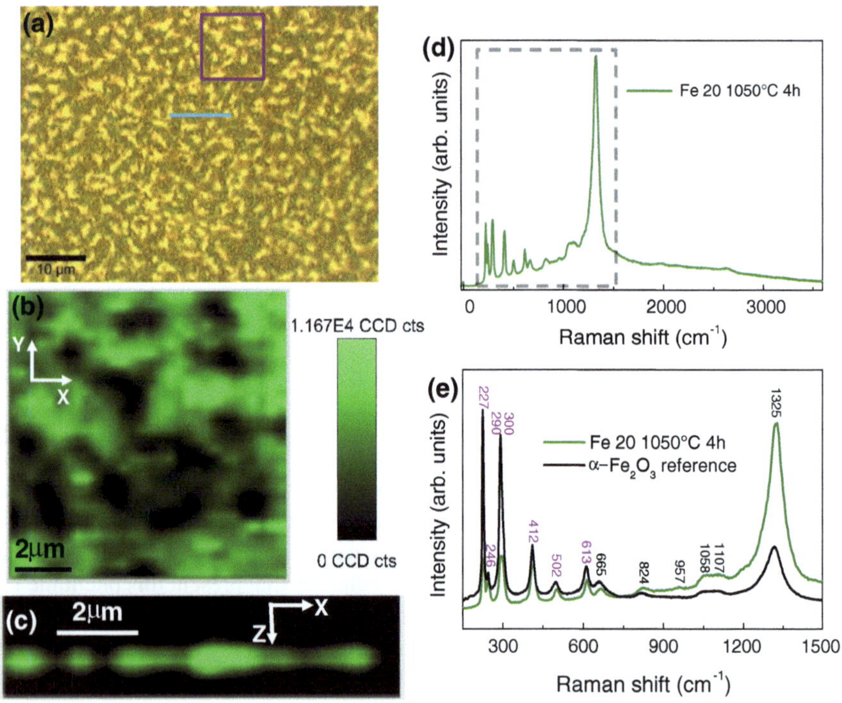

Fig. 4.10 Characterization by confocal Raman microscopy for Fe 20 1050 °C 4h: **a** optical micrograph, **b** in-plane and **c** in-depth Raman intensity image obtained from mapping the α-Fe_2O_3 Raman spectra taken each 300 nm and integrating in the spectral range from 0 to 3600 cm^{-1}, **d** average Raman spectrum obtained from in-plane Raman image and **e** a zoom of the average Raman spectrum from 150 to 1550 cm^{-1} compared with that measured by de Faria et al. [33]

in-depth Raman image shows dispersion effects at substrate/sample or sample/air interfaces. A magnification of the average Raman spectrum obtained from the in-plane Raman image (Fig. 4.10d), from 150 to 1550 cm^{-1}, is compared with the α-Fe_2O_3 spectrum measured by de Faria et al. [33], Fig. 4.10e. The Raman bands can be clearly associated with those of α-Fe_2O_3, identifying the seven phonon modes [33] allowed (purple numbers), besides others (black numbers) that are not allowed in Raman, but they have already been reported in the literature [34, 34–39]. This result agrees with those from Aggarwal et al. [18] who reported the fabrication of α-Fe_2O_3 NPs from an Fe film annealed at 1050 °C. Samples with initial Fe thicknesses of 5, 10 and 15 nm were also studied and Raman measurements showed qualitatively similar results to those obtained for sample Fe 20 1050 °C 4h.

- **Au/FeO$_x$ nanostructures**

Figures 4.11 and 4.12 show the Raman characterization carried out for a sample obtained from the first approach (Fe 20 1050 °C Au 10 500 °C, Fig. 4.4a) and the

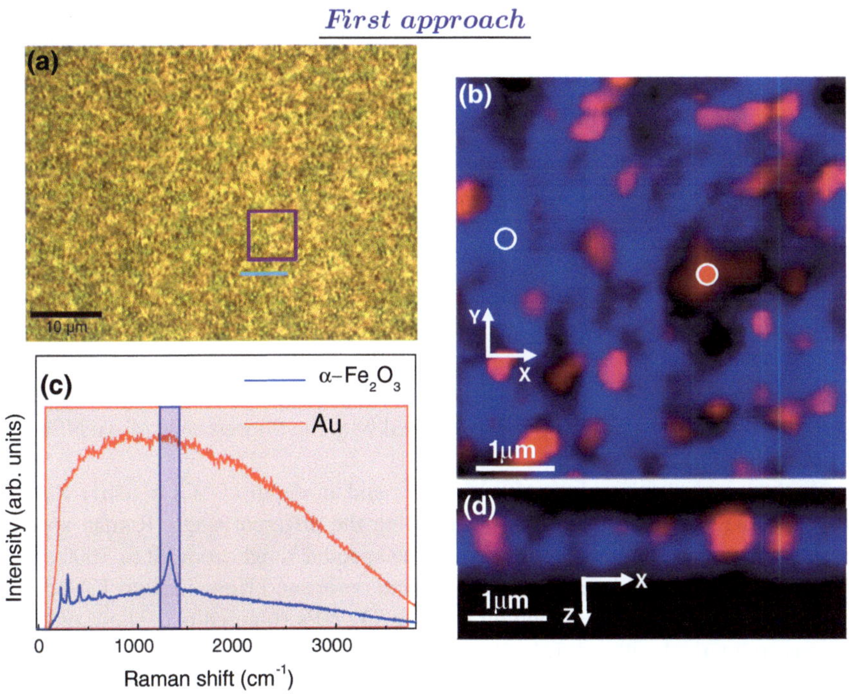

Fig. 4.11 Raman characterization for Fe 20 1050 °C Au 10 500 °C: **a** optical micrograph, **b** in-plane and **d** in-depth Raman intensity image obtained from mapping the different single Raman spectra taken each 100 nm and integrating in the spectral range from 90 to 3600 cm^{-1} for Au NPs (*red circle*) and from 1230 to 1380 cm^{-1} for α-Fe_2O_3 NPs (*blue circle*). **c** Single Raman spectra taken from in-plane Raman image, distinguishing the integrated regions (*red* region for Au NPs and *blue* region for α-Fe_2O_3 NPs) to obtain the Raman images

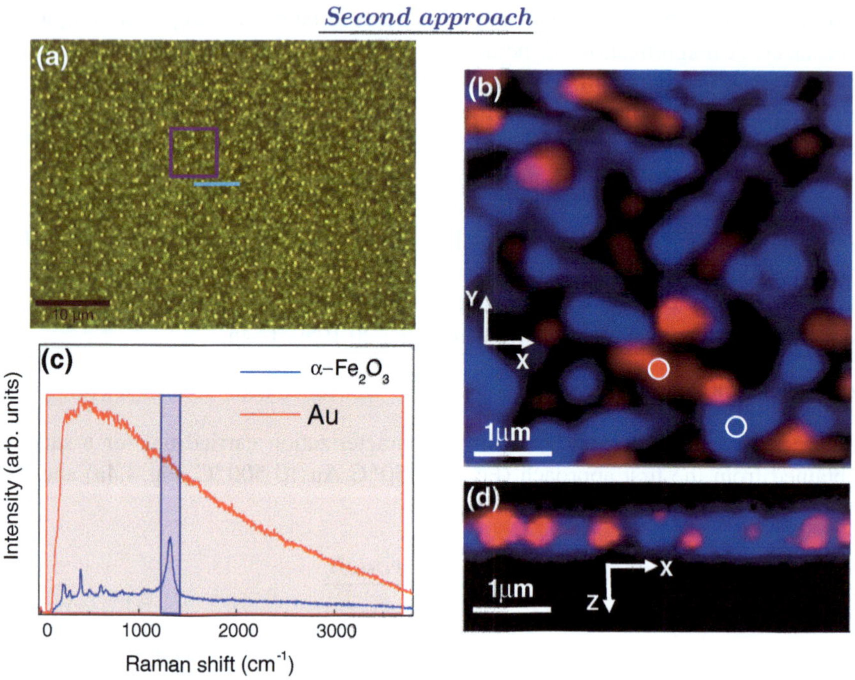

Fig. 4.12 Raman characterization for Au 10 Fe 5 1200 °C 30 min: **a** optical micrograph, **b** in-plane and **d** in-depth Raman intensity image obtained from mapping the different single Raman spectra taken each 100 nm and integrating in the spectral range from 90 to 3600 cm^{-1} for Au NPs (*red circle*) and from 1230 to 1380 cm^{-1} for α-Fe_2O_3 NPs (*blue circle*). **c** Single Raman spectra taken from in-plane Raman image, distinguishing the integrated regions (*red* region for Au NPs and *blue* region for α-Fe_2O_3 NPs) to obtain the Raman images

second approach (Au 10 Fe 5 1200 °C 30 min, Fig. 4.7c), respectively. Two types of Raman spectra were identified: one associated with the fluorescence of Au NPs and the other one related to that of α-Fe_2O_3 [33].

For both samples, in-plane (5 × 5 μm^2) and in-depth (5 × 1.5 μm^2) Raman intensity images were obtained from mapping the different single Raman spectra measured each 100 nm and integrating in the spectral range from 90 to 3600 cm^{-1} to distinguish the fluorescence of Au NPs (red color) and from 1230 to 1380 cm^{-1} to separate the signal corresponding to α-Fe_2O_3 NPs (blue color). These mapped regions are marked with a purple square (in-plane images) and a blue line (in-depth images) on the optical micrographs of the samples (see Fig. 4.11a for Fe 20 1050 °C Au 10 500 °C and Fig. 4.12a for Au 10 Fe 5 1200 °C 30 min). Single Raman spectra, shown in Figs. 4.11c and 4.12c, were taken from two regions marked with circles on the in-plane Raman images, corresponding to α-Fe_2O_3 (blue circles) and Au NPs (red circles) (see Figs. 4.11b and 4.12b). In these spectra, the regions that were integrated to obtain the Raman images are marked. The integration region chosen from 1230 to 1380 cm^{-1} for α-Fe_2O_3 NPs can be assigned to a second order scattering process about 1325 cm^{-1}, which corresponds to the band around 663 cm^{-1}, relative to the first order scattering [34]. This band can be associated with infrared (IR) active longitudinal optical (LO) E$_u$ mode and forbidden in Raman scattering (as detailed below), which is activated by disorder in the α-Fe_2O_3 crystals [35, 36].

Comparing the Raman characterization performed for both samples fabricated by the two different approaches in air, we can observe that Au NPs of the sample Fe 20 1050 °C Au 10 500 °C are more distributed on surface than those of the sample Au 10 Fe 5 1200 °C 30 min, where Au NPs are more agglomerated (Figs. 4.11b, d and 4.12b, d). In the case of the α-Fe_2O_3 NPs of sample Fe 20 1050 °C Au 10 500 °C, we obtain NPs distributed along the whole surface detecting few regions of the surface without covering. However, for the case of the sample Au 10 Fe 5 1200 °C 30 min, the α-Fe_2O_3 NPs are more agglomerated and they can be clearly identified, finding a higher number of regions on the surface without covering than in sample Fe 20 1050 °C Au 10 500 °C. These results are analogues to those obtained by the microstructural characterization (see Sect. 4.3.2). From in-depth Raman images (Figs. 4.11d and 4.12d), we can estimate the thickness of samples, obtaining values around 400–600 nm, larger than the real thicknesses. For these samples, the single Raman spectra to obtain the Raman images were taken each 100 nm. However, the optical resolution of the confocal microscope is of 500 nm in the vertical direction. This affects the Raman signal inducing dispersion phenomena at the substrate/sample or sample/air interfaces (similar to FeO$_x$ NPs, see Fig. 4.10).

All samples obtained by both approaches were measured by confocal Raman microscopy and in all cases, FeO$_x$ NPs were identified as α-Fe_2O_3 NPs and both the Au and α-Fe_2O_3 nanostructures were distinguished.

- **Analysis by Raman spectroscopy of FeO$_x$ nanostructures**

We analyze FeO$_x$ NPs of three samples fabricated by the different approaches with the same initial thicknesses of the Au and Fe films (Fig. 4.13): Au/FeO$_x$ NPs obtained by the first approach from a 20 nm Fe film annealed at 1050 °C for 4 h in air plus a 10 nm Au film annealed at 500 °C for 3 h in air (Fe 20 1050 °C Au 10 500 °C, Fig. 4.4c) and Au/FeO$_x$ NPs obtained by the second approach from a 10 nm Au/20 nm Fe bilayer annealed at 1200 °C for 3 and 30 min (Au 10 Fe 20 1200 °C 3 min and Au 10 Fe 20 1200 °C 30 min, Figs. 4.6g and 4.7g, respectively). In the Raman spectra, it is possible to distinguish the seven phonon modes ($2A_{1g} + 5E_g$) allowed in Raman for α-Fe_2O_3 (hematite), besides other bands. Around 663 cm^{-1}, IR active LO E$_u$ mode can be identified, which is forbidden in Raman scattering but is activated by disorder in the α-Fe_2O_3 crystals [34–36], as noted above. This band could be attributed to magnetite, which shows a peak around that wavenumber. However, high amounts of magnetite should be detected in the Raman spectrum and this is unlikely, since the complex NPs were obtained from thermal treatments at 1050 and 1200 °C, too high temperatures for obtaining magnetite. Therefore, the appearance in the Raman spectra of that vibrational mode may be explained by a non-perfect stoichiometry or by a small grain size, which increases the relative amount of surface atoms with different coordination spheres [36, 38], and thus the disorder. The vibrational mode at 820 cm^{-1} can be

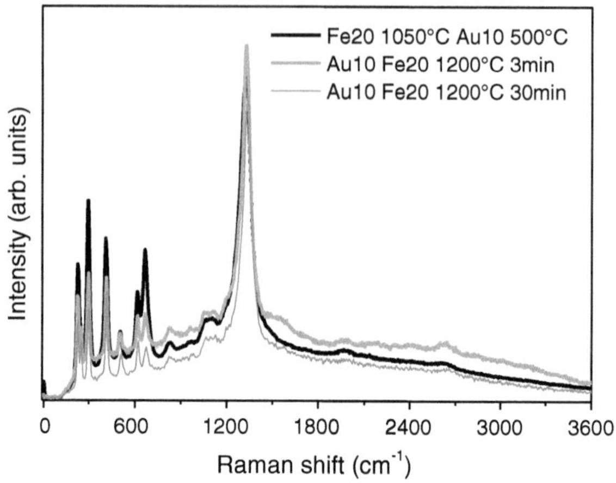

Fig. 4.13 Single Raman spectra for FeO$_x$ nanostructures of Fe 20 1050 °C Au 10 500 °C (*black line*), Au 10 Fe 20 1200 °C 3 min (red line) and Au 10 Fe 20 1200 °C 30 min (*blue line*). Spectra were taken with an integrating time of 2 s

assigned to a magnon. Hematite is an antiferromagnetic material with the spins slight canted above the Morin temperature, thus a collective spin movement (magnon) can be excited [37, 38]. Other modes around 950–1150 cm^{-1} may be due to the silica substrates, however their origin is not clear [36, 40]. The bands attributed up here correspond to fundamental phonon and magnon bands, however overtones of the first-order scattering can be observed. The band around 1325 cm^{-1} can be assigned to an overtone or second order scattering process of the band around 663 cm^{-1} ($2 \times 663 = 1326$ cm^{-1}), which is known to be strongly resonantly enhanced [34]. Moreover, at 1960 and 2630 cm^{-1}, the second ($3 \times 663 = 1989$ cm^{-1}) and third overtone ($4 \times 663 = 2652$ cm^{-1}) can be identified, respectively [34]. Finally, Raman scattering of two magnon mode may be identified at 1560 cm^{-1} ($2 \times 820 = 1640$ cm^{-1}) [39]. According to this assignation of Raman bands for α-Fe_2O_3 [33–42], Table 4.2 shows the wavenumbers obtained from a lorentzian fitting of Raman spectra for the three samples fabricated by the different approaches (see Fig. 4.13).

As evidenced Fig. 4.13, variations on the full width half maximum (FWHM), the Raman shift and relative intensities of the Raman modes can be observed. These modifications can be due to the morphological features of α-Fe_2O_3 NPs. Hence, in order to correlate the Raman signal with the size and the crystallinity of the nanostructures obtained from different annealing conditions, three Raman modes are analyzed: $E_g(5)$ and LO bands and the first overtone of the LO mode. The wavenumber, FWHM and intensity for the three Raman bands (Fig. 4.14) are presented in Table 4.3 for the three samples analyzed. Figure 4.14a illustrates the first overtone of the LO mode around 1325 cm^{-1}, where it is possible to observe that the peak is shifted towards smaller wavenumbers and the FWHM increases (the rest of the modes show a similar behavior) for NPs fabricated at lower temperatures by the first approach, which can be associated with a decrease in the size of α-Fe_2O_3 nanostructures [43]. For samples obtained by the second approach, we can observe that the first overtone of the LO mode for NPs annealed for 30 min results shifted towards larger wavenumbers with

Table 4.2 Raman modes and wavenumbers for α-Fe_2O_3 nanostructures of Fe 20 1050 °C Au 10 500 °C, Au 10 Fe 20 1200 °C 3 min and Au 10 Fe 20 1200 °C 30 min

Raman modes	Fe20 1050 °C Au10 500 °C ν (cm^{-1})	Au10 Fe20 1200 °C 3 min ν (cm^{-1})	Au10 Fe20 1200 °C 30 min ν (cm^{-1})
$A_{1g}(1)$	226.9	228.0	228.7
$E_g(1)$	246.8	247.3	250.8
$E_g(2)$	290.9	294.3	294.4
$E_g(3)$	300.2	300.9	302.8
$E_g(4)$	412.3	414.3	417.3
$A_{1g}(2)$	504.4	507.8	510.7
$E_g(5)$	612.9	614.8	619.8
Disorder	665.2	667.2	673.7
Magnon	827.8	830.1	831.1
	965.3	963.6	971.9
	1050.6	1051.2	1078.1
	1101.0	1118.3	1119.1
Overtone	1320.7	1325.8	1331.9
Two magnon	–	1558.9	–
Second overtone	1961.4	1967.5	1969.4
Third overtone	2628.8	2633.5	2643.6

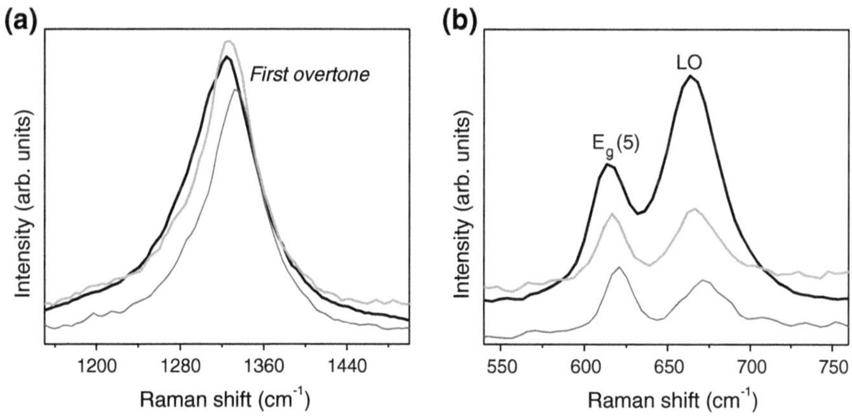

Fig. 4.14 **a** First overtone of the LO mode and **b** LO and $E_g(5)$ modes for Fe 20 1050 °C Au 10 500 °C (*black line*), Au 10 Fe 20 1200 °C 3 min (*red line*) and Au 10 Fe 20 1200 °C 30 min (*blue line*). Spectra were taken with an integrating time of 2 s

respect to those annealed for 3 min (the rest of the modes show a similar behavior), which can be related to an increase in the size of α-Fe_2O_3 nanostructures [43]. In general, wider Raman bands are observed for sample obtained by the first approach (see Table 4.3). These results suggest that NPs fabricated at higher temperatures and for

Table 4.3 Wavenumber, FWHM and intensity of three Raman modes for α-Fe_2O_3 nanostructures of samples: Fe 20 1050 °C Au 10 500 °C, Au 10 Fe 20 1200 °C 3 min and Au 10 Fe 20 1200 °C 30 min

	Fe20 1050 °C Au10 500 °C			Au10 Fe20 1200 °C 3 min			Au10 Fe20 1200 °C 30 min		
Raman modes	ν (cm^{-1})	FWHM (cm^{-1})	Intensity	ν (cm^{-1})	FWHM (cm^{-1})	Intensity	ν (cm^{-1})	FWHM (cm^{-1})	Intensity
$E_g(5)$	612.9	21.1	819.1	614.8	20.0	628.7	619.8	22.8	436.3
Disorder/LO	665.2	39.9	1138.7	667.2	33.9	644.6	673.7	43.4	380.5
Overtone	1320.7	78.0	2540.1	1325.8	61.2	2714.7	1331.9	60.9	2292.4

Table 4.4 Raman intensity ratio between the LO and $E_g(5)$ modes for α-Fe_2O_3 nanostructures of samples: Fe 20 1050 °C Au 10 500 °C, Au 10 Fe 20 1200 °C 3 min and Au 10 Fe 20 1200 °C 30 min

Intensity ratio	Fe20 1050 °C Au10 500 °C	Au10 Fe20 1200 °C 3 min	Au10 Fe20 1200 °C 30 min
$\frac{LO}{E_g(5)}$	1.39	1.02	0.87

larger annealing time present a higher crystallinity and larger NP size, which is consistent with the morphological characterization (see Sect. 4.3.2). On the other hand, the LO mode (Fig. 4.14b) can be related to the disorder in the α-Fe_2O_3 crystals [34–36]. A high contribution from the LO component with respect to other Raman modes can be attributed to a decrease in the crystallinity of hematite NPs [38]. In the three samples fabricated under different conditions, analyzing the intensity ratio for the LO and $E_g(5)$ modes (see Table 4.4), a larger lattice disorder is found for Fe 20 1050 °C Au 10 500 °C, where the annealing temperature is lower than in the other samples analyzed. For the samples obtained by the second approach at different annealing times (3 or 30 min), the sample treated for 30 min presents larger order than that annealed for 3 min. Hence, at low temperatures and short times of sintering, the crystalline disorder and the concentration of defects of nanostructures is higher, requiring high activation energy [41]. The lattice disorder can be also associated with the size of NPs, finding a high disorder for smaller size NPs, which show broader Raman bands and those shifted towards smaller wavenumbers. These effects increase as the nanocrystal size decreases, due to the phonon confinement [38].

Since Raman spectroscopy measures excitation of phonons along certain crystallographic direction, the intensity of Raman bands for each crystal depends on the relative orientation of the light polarization and lattice axis. Therefore, confocal Raman microscopy can provide intensity images distinguishing the different orientations of α-Fe_2O_3 NPs. Figure 4.15a shows in-plane Raman intensity image

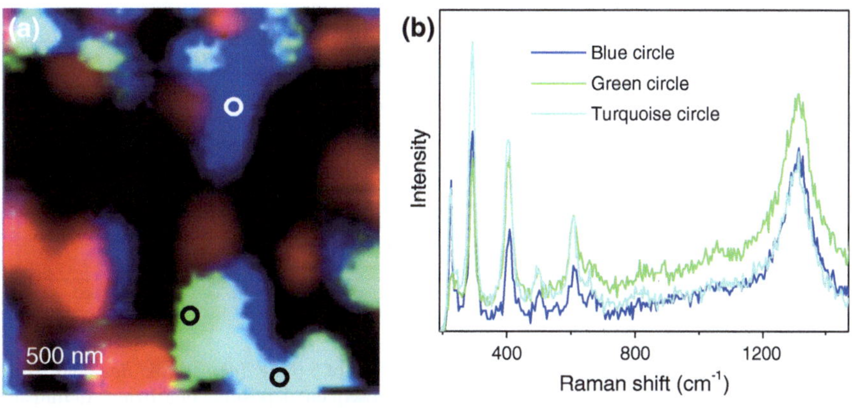

Fig. 4.15 **a** In-plane Raman intensity image obtained from mapping the different single Raman spectra taken each 50 nm and **b** single Raman spectra taken from three regions with different crystalline orientation of the α-Fe_2O_3 NPs for Au 10 Fe 20 1200 °C 3 min

(2.5 × 2.5 μm^2) for Au 10 Fe 20 1200 °C 3 min, which was obtained from mapping the different single Raman spectra measured each 50 nm and integrating in the spectral range from 90 to 3600 cm^{-1} to distinguish the Au NPs (red color) and from 266 to 312 cm^{-1} (blue color) and from 388 to 426 cm^{-1} (green color) to identify the α-Fe_2O_3 NPs with different crystalline orientation. The spectral ranges to discriminate the hematite signal were chosen because show Raman modes with different relative intensities. Single Raman spectra, shown in Fig. 4.15b, were taken from three points marked with circles on the in-plane Raman image that correspond to hematite NPs with different crystalline orientation.

4.3.4 Optical Characterization of Samples Prepared in Air

Optical absorption spectra of the samples prepared by annealing in air, shown in Fig. 4.16, were referenced to a silica substrate in order to obtain the net optical transitions of Au/α-Fe_2O_3 NPs.

Figure 4.16a shows the optical absorption spectra for three samples fabricated starting from an Fe thin film of 5, 10 and 20 nm annealed in air at 1050 °C for 4 h, plus a subsequent deposition of a 10 nm Au film and an annealing in air for 3 h at 500 °C (first approach). We can observe the surface plasmon (SP) band for the three samples, which appears between 700 and 1000 nm depending on Fe film thickness. The changes in the position of the absorption band related to SPs are due to the variations in the surrounding medium of the Au nanostructures. As explained in Chap. 1, the SP band shifts towards larger wavelengths as the dielectric permittivity of the surrounding medium increases [20]. The surrounding medium of the Au NPs is a mixture of air and iron oxide. Since the dielectric constant of FeO_x is larger than that of the air (for iron oxide is 2.3) [44], more abundance of iron oxide will induce

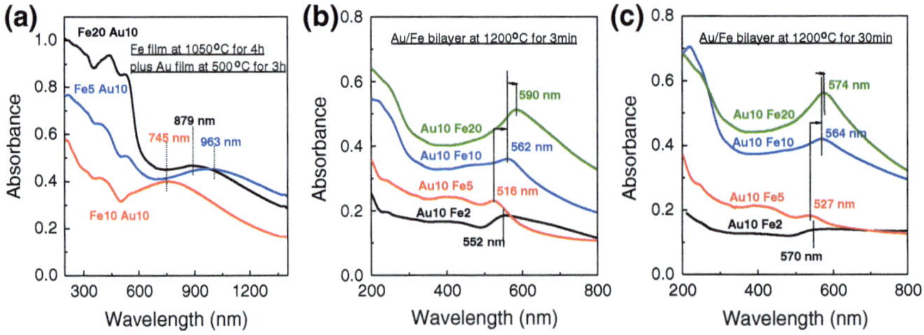

Fig. 4.16 **a** Optical absorption spectra for an Fe thin film of 5 (*blue line*), 10 (*red line*) and 20 nm (*black line*) annealed in air at 1050 °C for 4 h, plus a subsequent deposition of a 10 nm Au film and annealing in air for 3 h at 500 °C. **b** and **c** Optical absorption spectra for Fe/Au bilayers annealed at 1200 °C in air, with a 10 nm Au thickness and varying the Fe film thickness: 2 (*black line*), 5 (*red line*), 10 (*blue line*) and 20 nm (*green line*), for 3 and 30 min, respectively

a larger shift of the SP band. However, sample with a initial Fe film of 5 nm presents the SP band at 963 nm, a wavelength larger than for samples with a initial Fe film of 10 and 20 nm. Therefore, we have to consider that SP position is also modified by the different morphology of Au NPs. This interpretation is consistent with the SEM images (see Fig. 4.2), where we observed a short inter-particle distance in the sample Fe 5 1050 °C Au 10 500 °C, which favors the dipolar interaction between adjacent particles and is known to widen and red-shift the SP band [1, 45], as Table 4.5 shows. On the other hand, optical absorption spectra for samples consisting of 10 nm Fe/Au bilayers with different Fe thicknesses ranging from 2 to 20 nm annealed at 1200 °C for 3 and 30 min are shown in Fig. 4.16b, c, respectively (second approach). For this second approach, the absorption bands associated with the SPs of Au NPs also shifts with the initial Fe thickness and they are clearly observed between 500 and 600 nm. The plasmon resonance shifts towards larger wavelengths when the initial Fe layer thickness increases, except for 2 nm Fe/10 nm Au bilayers, probably due to size effects. This shift of the SP band can be explained because of the variations in the dielectric permittivity of the surrounding medium of the Au NPs, as indicated above. The formation of complex nanostructures modify the optical properties of Au NPs due to proximity effects. Hence, the morphological features of $Au/\alpha\text{-}Fe_2O_3$ NPs can also affect to the SP position.

The SP width can be related to a high size dispersion of Au NPs (see morphological characterization in Sect. 4.3.2). Samples fabricated by the first approach show SPs with the FWHM larger than for samples prepared by the second approach, as Table 4.5 shows. This can be related to Au NPs obtained by the first approach present a higher size dispersion. Within the first approach, we can observe that sample Fe 10 1050 °C Au 10 500 °C shows the resonance band intensity higher and the FWHM narrower than for samples with an Fe film thickness of 5 and 20 nm (see Table 4.5). This could be related to larger Au NPs for sample with an Fe film thickness of 10 nm.

Table 4.5 Parameters corresponding to the gaussian fits of the SP bands for samples prepared by both approaches in air

Sample	Wavelength (nm)	FWHM (nm)	Intensity
Fe 5 1050 °C Au 10 500 °C	963	610	0.06
Fe 10 1050 °C Au 10 500 °C	745	382	0.11
Fe 20 1050 °C Au 10 500 °C	879	453	0.07
Au 10 Fe 2 1200 °C 3 min	552	181	0.04
Au 10 Fe 5 1200 °C 3 min	516	128	0.05
Au 10 Fe 10 1200 °C 3 min	562	143	0.06
Au 10 Fe 20 1200 °C 3 min	590	159	0.10
Au 10 Fe 2 1200 °C 30 min	570	110	0.02
Au 10 Fe 5 1200 °C 30 min	527	71	0.05
Au 10 Fe 10 1200 °C 30 min	564	125	0.05
Au 10 Fe 20 1200 °C 30 min	574	119	0.11

For samples obtained by the second approach, the SP intensity is higher for samples prepared starting from a 20 nm Fe film, which can be related to larger Au NPs and the increase of the dielectric function of the medium. The lowest SP intensity is for the samples with an Fe film thickness of 2 nm, suggesting smaller Au NPs. These results agree with the morphological characterization (Sect. 4.3.2).

On the other hand, absorption bands that correspond to the electronic transitions of iron oxide can be distinguished and identified according to the literature [21, 46]. The absorption bands in the region between 200 and 400 nm result from ligand to metal charge transfer transitions and partly from the contributions of Fe^{3+} ligand field transitions $^6A_1 \rightarrow ^4 T_1(^4P)$ at 290–310 nm, $^6A_1 \rightarrow ^4 E(^4D)$ and $^6A_1 \rightarrow ^4 T_2(^4D)$ at 360–380 nm. For the region between 400 and 600 nm, the absorption bands correspond to the pair excitation processes $^6A_1 +^6 A_1 \rightarrow ^4 T_1(^4G) +^4 T_1(^4G)$ at 485–550 nm and the contributions of $^6A_1 \rightarrow ^4 E, ^4 A_1(^4G)$ ligand field transitions around 430–440 nm. The transition at 640 nm assigned to the $^6A_1 \rightarrow ^4 T_2(^4G)$ cannot be resolved because it is masked by the bands associated with SPs, for all samples. Moreover, for samples obtained by the second approach, it is possible to observe absorption bands related to the transitions of iron oxide NPs less intense than those measured for NPs obtained by the first approach. Samples with more surface without covering by NPs show lower absorption (see morphological features in Sect. 4.3.2).

In summary, for both annealing approaches, we can distinguish the absorption band associated with the SPs of Au nanostructures, which can be related to the initial Fe thickness that is the only modified parameter. The optical measurements are a direct confirmation of the coupling of both kinds of NPs (Au and α-Fe_2O_3 NPs) evidenced by the AFM images (see Figs. 4.3, 4.6, 4.7, 4.8 and 4.9).

4.4 Au/FeO$_x$ Nanostructures Obtained by Annealing in Low Vacuum

4.4.1 Fabrication of Samples Prepared in Low Vacuum

The single iron oxide phase that we have been able to obtain by means of thermal treatments in air has been α-Fe_2O_3, which exhibits an antiferromagnetic character, different to γ-Fe_2O_3 or Fe_3O_4, which are ferrimagnetic materials, showing consequently additional properties. Particularly, γ-Fe_2O_3 nanostructures have a great interest because of their magnetic behavior [21] and their fabrication by means of thermal treatments of thin films have not been reported up to our knowledge. Besides, the fabrication of γ-Fe_2O_3 NPs in combination with Au NPs results appealing since they can exhibit coupling between the magnetic and optical properties of NPs [23] due to surface or proximity effects.

It is well known that α-Fe_2O_3 can be reduced to Fe_3O_4 by annealing in nitrogen at 500 °C [18]. However, our samples are constituted by α-Fe_2O_3 and Au NPs, thus a post-annealing at 500 °C in nitrogen may induce the transformation of α-Fe_2O_3 to

Fe_3O_4 but structural changes on Au NPs as well. Moreover, the fabrication of complex NPs in just one step is more advantageous. As mentioned above, annealing of samples in high vacuum or reducing atmospheres in order to prevent iron oxidation is not a chance. Therefore, in order to fabricate nanostructures constituted by Au and FeO$_x$ NPs, where the iron oxide is different to α-Fe_2O_3, we carried out thermal treatments of Fe/Au bilayers grown on silica substrates at 1100 °C for 12 h in low vacuum, around 10^{-3} torr. During heating and cooling steps of annealing, the temperature increased and decreased 5 °C/min until or from the target temperature. Since for thermal treatments in low vacuum the oxygen concentration is lower than in air, the annealing time was increased at 12 h in order to favor surface diffusion [29, 47]. In the literature, there are few studies where NPs are fabricated by annealing in vacuum. Therefore, in order to compare the results of complex NPs with those of the individual components, we prepared samples with the configuration shown in Fig. 4.17. First, a 10 nm Au film was grown on a silica substrate, leaving an uncovered substrate region. Later, an Fe film varying the thickness (5 and 10 nm) was evaporated covering the bare substrate and leaving an uncovered Au film region. In this way, Au, FeO$_x$ and Au/FeO$_x$ NPs on the same silica substrate and under the same annealing conditions were obtained. The area of the substrate that was coated with the metallic films was of 2×2 cm^2. Table 4.6 shows a short description of the Au/FeO$_x$ structures obtained.

It should be noted that we repeated the experiment twice with nominally the same annealing conditions. However, the results were different in both cases. At the first time Au NPs were obtained while at the second time not. Besides, the FeO$_x$ and Au/FeO$_x$ structures seemed more sintered at the second attempt. The experiments were repeated several times and the results were always similar to those obtained at the second time. We could not obtain the same results as at the first time. Since the morphology of nanostructures obtained by both annealing attempts and their optical properties were quite interesting, we show below the results for the different samples

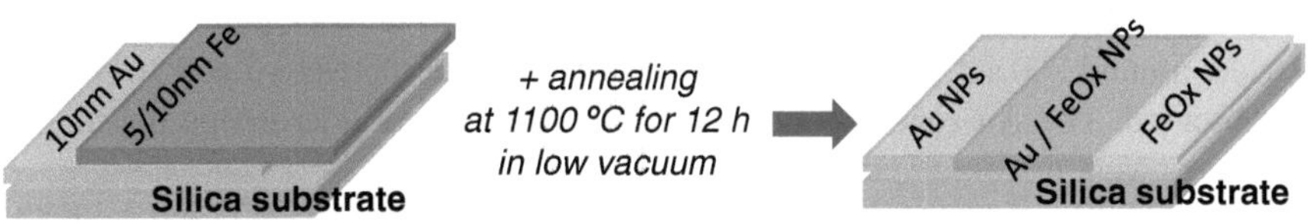

Fig. 4.17 Sample configuration to fabricate Au, FeO$_x$ and Au/FeO$_x$ NPs in low vacuum at 1100 °C for 12 h

Table 4.6 Samples fabricated from Fe/Au bilayers at two attempts in low vacuum

Sample	Description
First attempt	
Fe 10 Au 10 1100 °C 12 h_v_1	10 nm Fe/10 nm Au bilayer annealed at 1100 °C for 12 h
Second attempt	
Fe 10 Au 10 1100 °C 12 h_v_2	10 nm Fe/10 nm Au bilayer annealed at 1100 °C for 12 h
Fe 5 Au 10 1100 °C 12 h_v_2	5 nm Fe/10 nm Au bilayer annealed at 1100 °C for 12 h

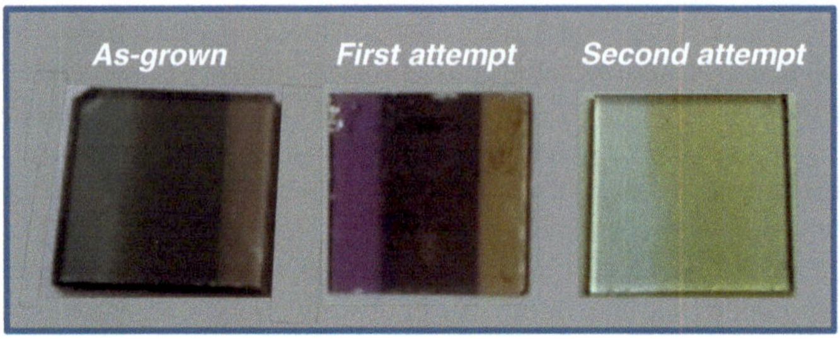

Fig. 4.18 Photographs of sample grown from 10 nm Fe and 10 nm Au and further annealed at 1100 °C for 12 h in low vacuum at the first and second attempt

fabricated. Figure 4.18 shows samples prepared by this process at 1100 °C for 12 h at both the first and second attempt, together with the as-grown sample from 10 nm Fe and 10 nm Au.

4.4.2 Morphological Characterization of Samples Prepared in Low Vacuum

- **First attempt**

For the sample obtained at the first attempt in low vacuum, sample Fe 10 Au 10 1100 °C 12 h_v_1, we could differentiate three zones that correspond to Au, FeO_x and Au/FeO_x zone. Figure 4.19 shows the optical microscope images for FeO_x and Au/FeO_x zones, where it is possible to observe structures with sizes of the order of several microns. Specifically, for the FeO_x zone, two kinds of structures can be distinguished: structures around 3–5 μm and microwires with a length of 3 μm and a width around 500 nm. For the Au/FeO*x* zone, structures between 1 and 10 μm can be observed. The Au zone was also studied by the optical microscope but the Au structures were not observed, likely their size was too small.

Figures 4.20 and 4.21 show the AFM images corresponding to FeO_x and Au/FeO_x zones of the sample Fe 10 Au 10 1100 °C 12 h_v_1, respectively. In the case of FeO_x zone, structures show a height around 10 nm and these are nanostructured by NPs with a size of the order of 40 nm and a height between 3 and 6 nm. It should be noted that the microwires observed by optical microscopy in this zone (see Fig. 4.19) could not be observed by AFM. On the other hand, AFM images for Au/FeO_x zone show microstructures with a height around 10 nm, which are formed by NPs with a size of 30–50 nm and a height around 2–5 nm. NPs with a size between 50 and 100 nm and a height of the order of 20 nm are also observed, finding them more clustered close to the microstructures (see Fig. 4.21c).

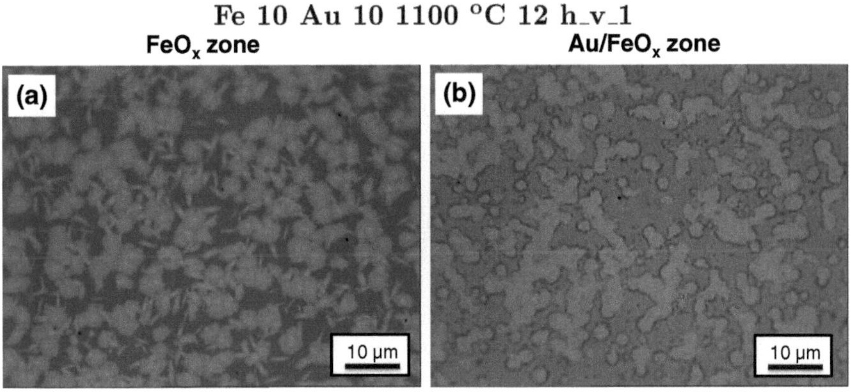

Fig. 4.19 Optical microscope images for **a** FeO$_x$ and **b** Au/FeO$_x$ zones of sample Fe 10 Au 10 1100 °C 12 h_v_1 obtained from the first attempt

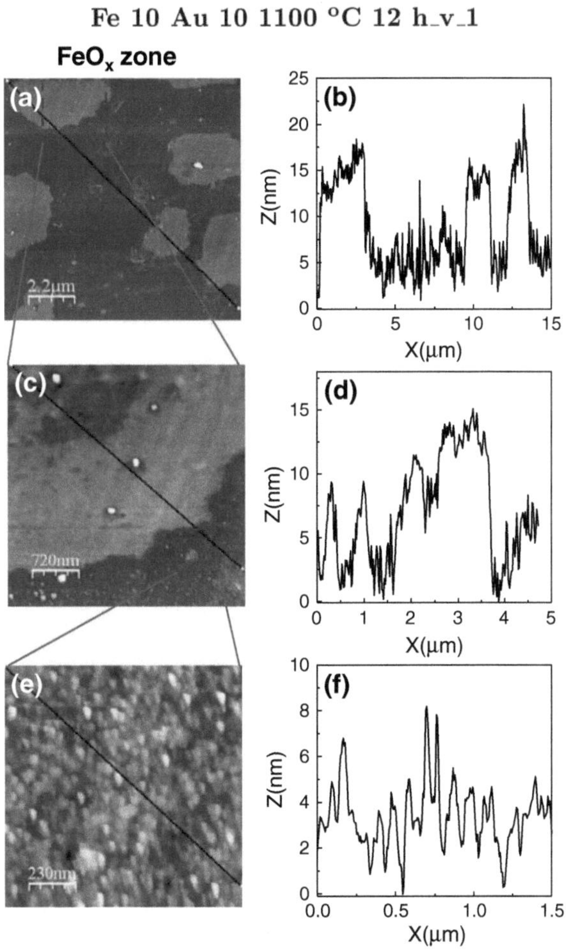

Fig. 4.20 Topography AFM images for the FeO$_x$ zone of sample annealed at the first attempt, from 10 nm Fe and 10 nm Au grown on a silica substrate, at 1100 °C for 12 h in low vacuum, around 10^{-3} torr. *Right panels* present height profiles measured along the lines indicated on AFM images

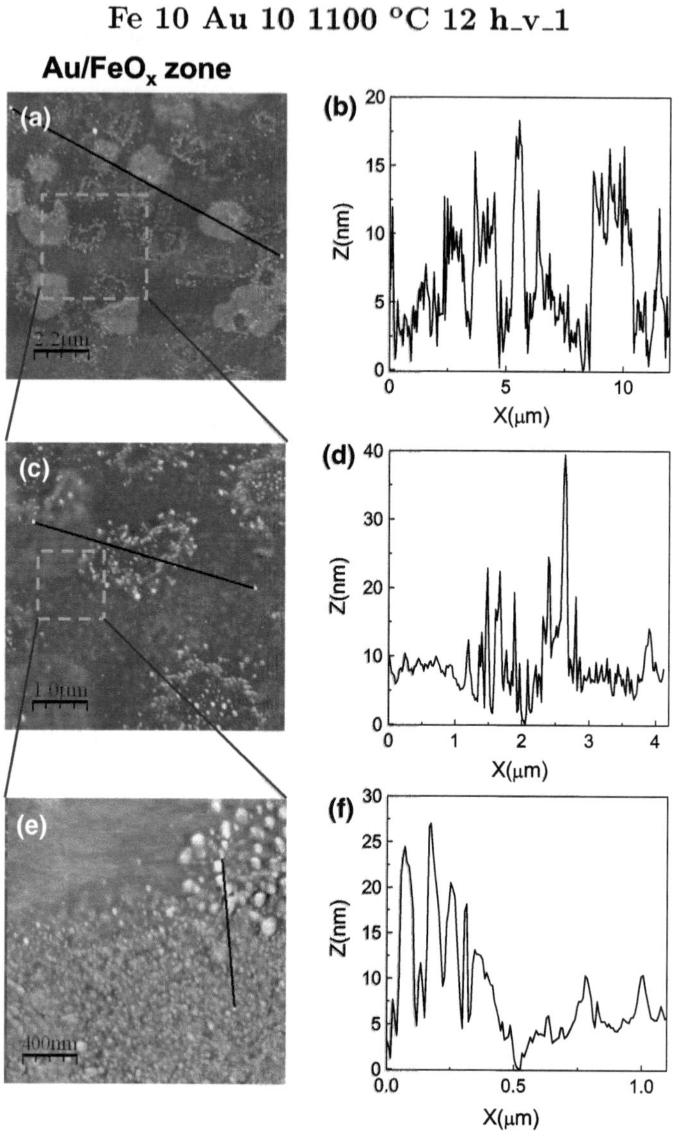

Fig. 4.21 Topography AFM images for the Au/FeO$_x$ zone of sample annealed at the first attempt, from 10 nm Fe and 10 nm Au grown on a silica substrate, at 1100 °C for 12 h in low vacuum, around 10^{-3} torr. *Right panels* present height profiles measured along the lines indicated on AFM images

- **Second attempt**

As mentioned above, samples prepared at the second attempt under same nominal conditions as at the first attempt, showed different morphological results. In this case, samples varying the Fe film thickness, 5 and 10 nm, and fixing the Au film thickness to 10 nm were obtained, samples Fe 5 Au 10 1100 °C 12 h_v_2 and Fe 10 Au 10 1100 °C 12 h_v_2 respectively. They presented two zones that correspond to FeO_x and Au/FeO_x zone, where structures are distributed homogeneously over the whole substrate surface, as shown SEM images (Fig. 4.22). In both zones of sample Fe 10 Au 10 1100 °C 12 h_v_2 (Fig. 4.22a, b), structures present an average size below 1 μm. For the sample Fe 5 Au 10 1100 °C 12 h_v_2 (Fig. 4.22c, d), structures with a size below 500 nm can be differentiated. With this attempt, structures for both samples in the Au zone were not obtained. The Au was evaporated from the substrate during annealing.

AFM measurements for the sample Fe 10 Au 10 1100 °C 12 h_v_2 were carried out in order to compare with the sample obtained at the first attempt. For FeO_x zone, Fig. 4.23, nanostructures can be distinguished with a height of 5–30 nm and a lateral size from 100 to 800 nm. Moreover, a detail of the AFM image shows that

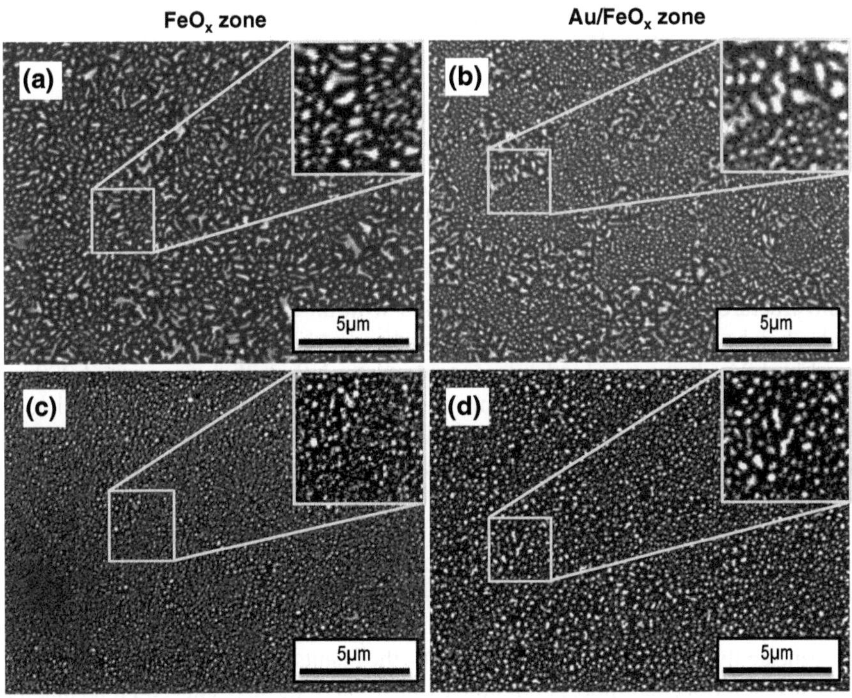

Fig. 4.22 SEM images for **a** and **c** FeO_x and **b** and **d** Au/FeO_x zones of samples Fe 10 Au 10 1100 °C 12 h_v_2 (**a** and **b**) and Fe 5 Au 10 1100 °C 12 h_v_2 (**c** and **d**), obtained in low vacuum at the second attempt

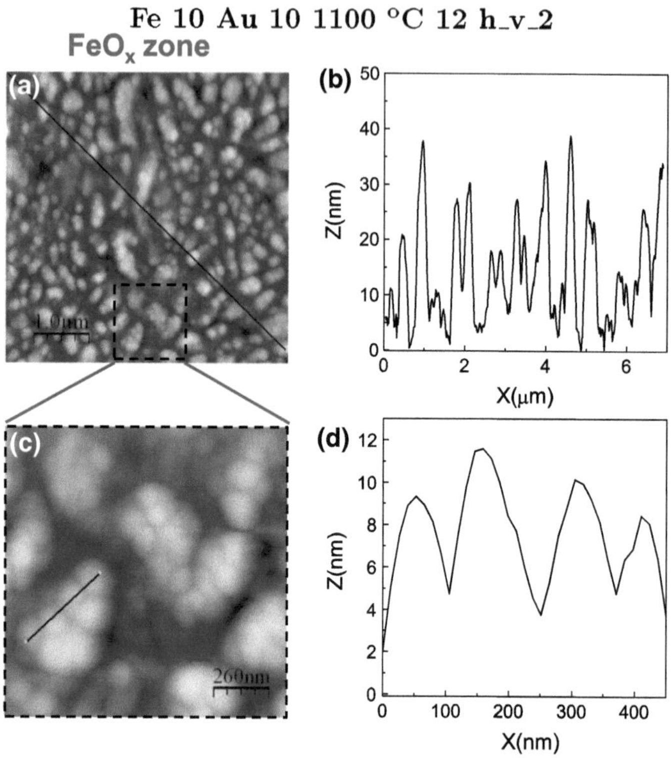

Fig. 4.23 Topography AFM images for the FeO_x zone of sample obtained at the second attempt from 10 nm Fe and 10 nm Au grown on a silica substrate and annealed at 1100 °C for 12 h in low vacuum. *Right panels* present height profiles measured along the lines indicated on AFM images

nanostructures are formed by rounded NPs with an average size of 100–150 nm and a height around 4–8 nm. On the other hand, AFM image for Au/FeO_x zone, see Fig. 4.24, shows two kinds of nanostructures: NPs with a size around 200 nm and others with a size between 500 nm and 1 μm can be differentiated, showing an average height similar in both kinds of NPs, around 10–40 nm. Sample Fe 5 Au 10 1100 °C 12 h_v_2 was also studied by AFM and the results were qualitatively similar to those of sample Fe 10 Au 10 1100 °C 12 h_v_2.

For samples obtained at this second attempt, it is possible to observe more sintered FeO_x and Au/FeO_x structures than those fabricated at the first attempt. Besides, single Au NPs were only obtained for the sample fabricated at the first attempt (the substrate looked colored in the Au zone, Fig. 4.18). The experiments were carried out repeatedly and at the following attempts, the results were always similar to those obtained at the second time. A possible explication of these results could be that the vacuum during annealing changed from the first to second time, acquiring a higher pressure. However, as shown with the Raman characterization (see Sect. 4.4.3), a

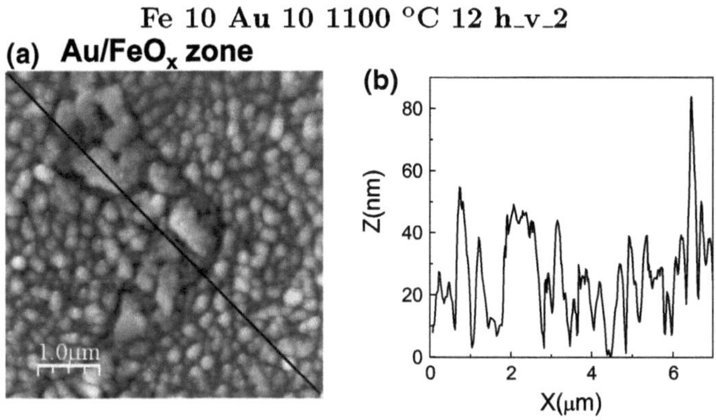

Fig. 4.24 **a** Topography AFM image for the Au/FeO$_x$ zone of sample obtained at the second attempt from 10 nm Fe and 10 nm Au grown on a silica substrate and annealed at 1100 °C for 12 h in low vacuum. **b** Height profile measured along the lines indicated on AFM image

larger amount of γ-Fe_2O_3 than α-Fe_2O_3 is obtained in the sample prepared at the second attempt, suggesting a lower oxygen proportion. Therefore, the process is not clear.

4.4.3 Confocal Raman Characterization of Samples Prepared in Low Vacuum

Samples fabricated in low vacuum were also studied by confocal Raman microscopy using the same procedure than for the samples fabricated in air.

- **First attempt**

Figure 4.25 shows the Raman characterization for the FeO$_x$ zone of the sample Fe 10 Au 10 1100 °C 12 h_v_1, in which we identified two types of Raman spectra: one associated with γ-Fe_2O_3 [48] and the other one related to that of α-Fe_2O_3 [33] (see Fig. 4.25d). In-plane (16 × 16 μm^2) and in-depth (8 × 2 μm^2) Raman intensity images were obtained, which are marked with a blue line (in-depth image) and a purple square (in-plane image) on the optical micrograph of the sample (see Fig. 4.25a). Raman spectra, shown in Fig. 4.25d, represent the average spectra corresponding to each phase of iron oxide obtained from the in-plane Raman image. Figure 4.25e shows a zoom of the Raman spectra from 0 to 1550 cm^{-1}, identifying the Raman bands for α-Fe_2O_3 (green numbers) and γ-Fe_2O_3 (blue numbers) nanostructures [33, 48]. For this zone, a larger amount of α-Fe_2O_3 than of γ-Fe_2O_3 was obtained, around 70 %.

For the case of the Au/FeO$_x$ zone of the sample Fe 10 Au 10 1100 °C 12 h_v_1, in-depth Raman intensity image (10 × 2 μm^2) was measured (Fig. 4.26b) and is marked with a blue line on the optical micrograph of the sample (see Fig. 4.26a). The Raman

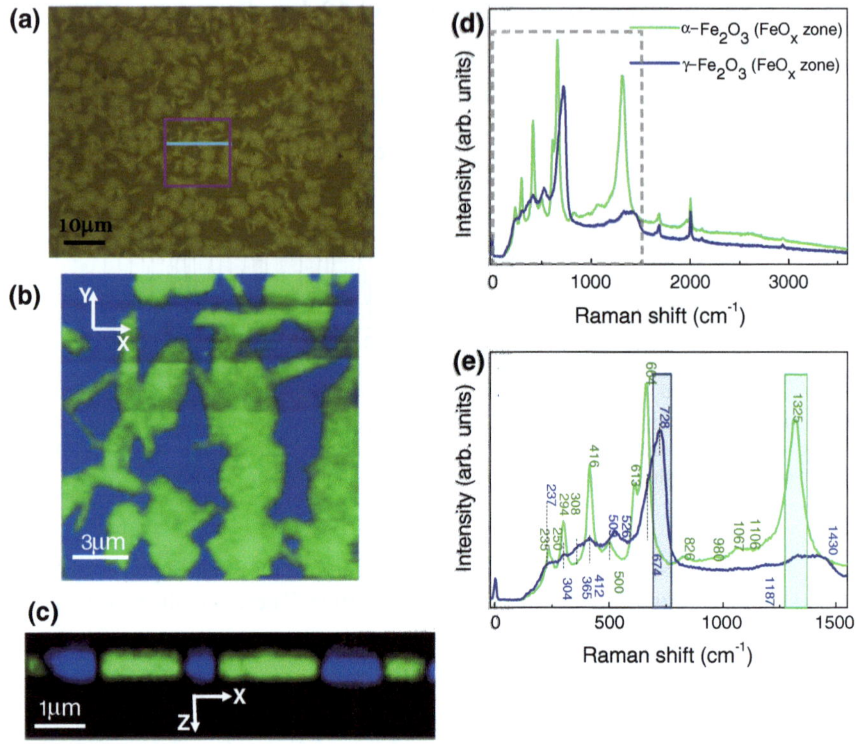

Fig. 4.25 Raman characterization for the FeO$_x$ zone of the sample Fe 10 Au 10 1100 °C 12 h_v_1: **a** optical micrograph, **b** in-plane and **c** in-depth Raman intensity images obtained from mapping the different single Raman spectra taken each 200 nm and integrating in the spectral range from 700 to 760 cm^{-1} for γ-Fe_2O_3 and from 1260 to 1360 cm^{-1} for α-Fe_2O_3, **d** average Raman spectra for each phase of iron oxide from in-plane Raman image and **e** a detail of the spectra from 0 to 1550 cm^{-1}, identifying the Raman modes and distinguishing the integrated regions (*green* region for α-Fe_2O_3 and *blue* region for γ-Fe_2O_3 NPs) to obtain the Raman images

image reveals the high Au fluorescence that masks the rest of signals that correspond to the iron oxides. This implies a complex distinction between the different species of iron oxide. For this case, single Raman spectra, shown in Fig. 4.26c, were taken on the in-depth Raman intensity image instead of on the in-plane image (as for the Raman characterization of the rest of the samples), which could not be obtained due to the high fluorescence. This high Au fluorescence did not allow identifying clearly the iron oxide spectra. Thus, Raman spectra corresponding to α-Fe_2O_3 and γ-Fe_2O_3 obtained from FeO$_x$ zone for the same sample (see Fig. 4.25d) are also shown, in order to identify the bands related to the different species of iron oxide. Comparing spectra, we detect α-Fe_2O_3, γ-Fe_2O_3 and Au NPs (see Fig. 4.26b, c).

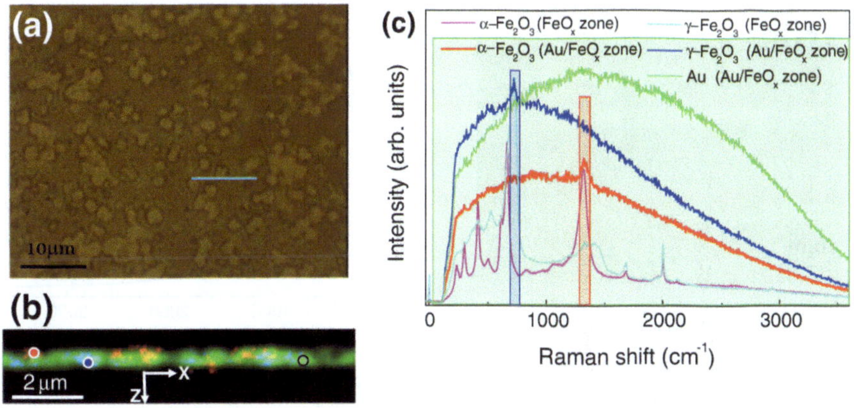

Fig. 4.26 Raman characterization for the Au/FeO$_x$ zone of the sample Fe 10 Au 10 1100 °C 12 h_v_1: **a** optical micrograph, **b** in-depth Raman intensity image obtained from mapping the different single Raman spectra taken each 100 nm and integrating in the spectral range from 700 to 760 cm^{-1} for γ-Fe_2O_3, from 1260 to 1360 cm^{-1} for α-Fe_2O_3 and from 90 to 3600 cm^{-1} for the Au nanostructures, and **c** single Raman spectra taken from Raman image and those of α-Fe_2O_3 and γ-Fe_2O_3 correspond to FeO$_x$ zone in order to distinguish the iron oxide phases. Integrated regions (*blue* region for γ-Fe_2O_3, *red* region for α-Fe_2O_3 and *green* region for the Au NPs) to obtain the Raman image are shown

- **Second attempt**

Figure 4.27 shows the Raman characterization for the FeO$_x$ zone of the sample obtained at the second attempt, Fe 10 Au 10 1100 °C 12 h_v_2. For this region, we distinguished the Raman spectra associated with γ-Fe_2O_3 [48] and α-Fe_2O_3 [33], similar to the first attempt. In-plane (10 × 10 μm^2) and in-depth (16 × 4 μm^2) Raman intensity images were obtained and are marked with a blue line (in-depth image) and a purple square (in-plane image) on the optical micrograph of the sample (see Fig. 4.27a). For the whole FeO$_x$ zone we identified mainly γ-Fe_2O_3 NPs, except some areas that are related to α-Fe_2O_3 NPs [33]. Average Raman spectra taken from the in-plane Raman image for each iron oxide are illustrated in Fig. 4.27d, and an identification of Raman bands is shown in Fig. 4.27e. In this case, the γ-Fe_2O_3 proportion is of the order of 99 %.

Finally, Raman characterization for Au/FeO$_x$ zone of the sample Fe 10 Au 10 1100 °C 12 h_v_2 is presented in Fig. 4.28. For this zone, it is possible to separate clearly the Raman spectra corresponding to α-Fe_2O_3, γ-Fe_2O_3 and that associated with the fluorescence of the Au nanostructures. In-plane (8 × 8 μm^2) and in-depth (7.5 × 2 μm^2) Raman intensity images were obtained and the mapped areas are marked on the optical micrograph of the sample (see Fig. 4.28a). Single Raman spectra of α-Fe_2O_3, γ-Fe_2O_3 and Au NPs, shown in Fig. 4.28d, were taken from three points marked on the in-plane Raman image. Around 80 % of iron oxide in γ phase is obtained.

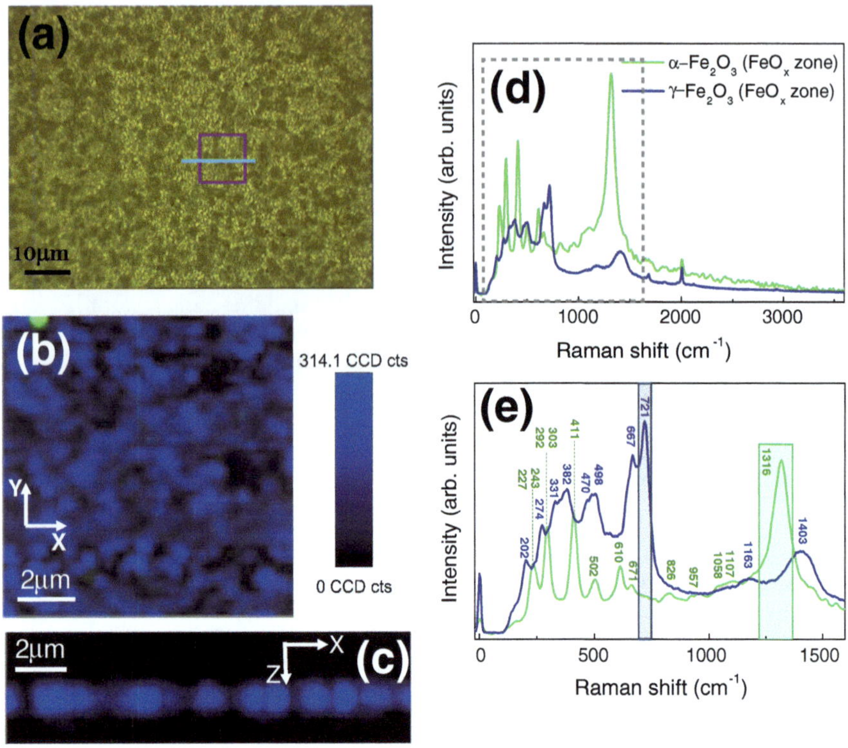

Fig. 4.27 Raman characterization for the FeO$_x$ zone of the sample Fe 10 Au 10 1100 °C 12 h_v_2: **a** optical micrograph, **b** in-plane and **c** in-depth Raman intensity images obtained from mapping the different single Raman spectra taken each 200 nm and integrating in the spectral range from 690 to 760 cm^{-1} for γ-Fe_2O_3 and from 1260 to 1390 cm^{-1} for α-Fe_2O_3, **d** average Raman spectra of in-plane Raman image and **e** a zoom of the average Raman spectrum from 0 to 1550 cm^{-1}, identifying the Raman modes and distinguishing the integrated regions (*green* region for α-Fe_2O_3 and *blue* region for γ-Fe_2O_3 NPs) to obtain the Raman images

For both attempts, Raman spectra of FeO$_x$ structures are associated to those of α-Fe_2O_3 and γ-Fe_2O_3 [33, 48]. In order to represent the intensity Raman images, the integration region chosen around 1220–1390 cm^{-1} for α-Fe_2O_3 NPs was the same than for the samples fabricated in air. As discussed above, this Raman band can be assigned to a second order scattering process about 1320 cm^{-1}, which corresponds to the observed band around 665 cm^{-1} (IR active LO E_u mode) relative to the first order scattering [34–36]. On the other hand, the integrated region around 690–780 cm^{-1} for γ-Fe_2O_3 NPs can be assigned to a first order scattering process, about 725 cm^{-1}, which corresponds to an A_{1g} mode that is allowed in the Raman scattering [43, 48].

With these results obtained by confocal Raman microscopy for samples fabricated from 10 nm Fe and 10 nm Au grown on silica substrate and annealed at 1100 °C for 12 h in low vacuum, we have been able to distinguish the iron oxide and Au nanostructures present in the samples, except for the Au/FeO$_x$ zone of the sample Fe 10

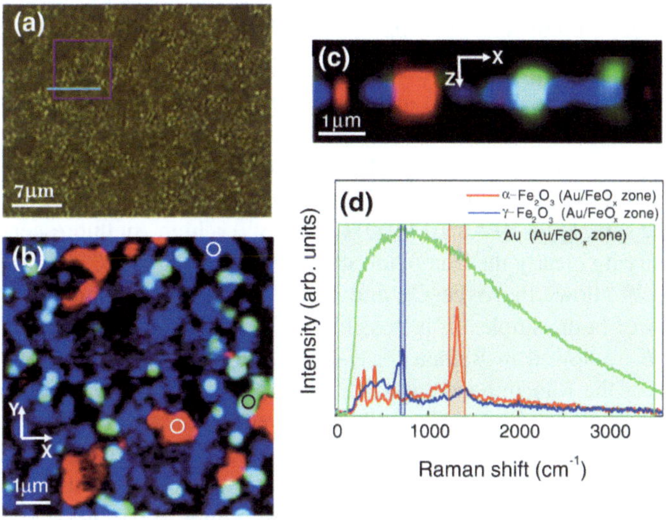

Fig. 4.28 Raman characterization for the Au/FeO$_x$ zone of the sample Fe 10 Au 10 1100 °C 12 h_v_2: **a** optical micrograph, **b** in-plane and **c** in-depth Raman intensity images obtained from mapping the different single Raman spectra taken each 100 nm and integrating in the spectral range from 690 to 780 cm^{-1} for γ-Fe_2O_3, from 1220 to 1390 cm^{-1} for α-Fe_2O_3 and from 70 to 3600 cm^{-1} for the Au nanostructures and **d** single Raman spectra taken from in-plane Raman image, distinguishing the integrated regions (*blue* region for γ-Fe_2O_3, *red* region for α-Fe_2O_3 and *green* region for the Au NPs) to obtain the Raman images

Au 10 1100 °C 12 h_v_1 where a clear distinction of nanostructures was not possible due to high fluorescence of Au NPs. From in-depth Raman images (Figs. 4.25c, 4.26b, 4.27c and 4.28c), we can estimate the thickness of samples, obtaining values around 600–1000 nm. For FeO$_x$ and Au/FeO$_x$ zones, the single Raman spectra to obtain the Raman images were taken each 200 nm and 100 nm, respectively. However, as explained above, the optical resolution of the confocal microscope in the vertical direction is of 500 nm. Thus, this resolution limit determines that the thickness values are larger than the real thicknesses. A Raman characterization for the sample obtained at the second attempt starting from a 5 nm Fe film, sample Fe 5 Au 10 1100 °C 12 h_v_2, was also analyzed. For Au/FeO$_x$ zone, we found α-Fe_2O_3, γ-Fe_2O_3 and Au NPs with a proportion of around 60 % of iron oxide in γ phase. This percentage of γ-Fe_2O_3 is a 20 % smaller than for sample Fe 10 Au 10 1100 °C 12 h_v_2, in which the initial thickness of Fe film was larger (10 nm instead of 5 nm). Besides, a larger amount of uncovered substrate than for sample Fe 10 Au 10 1100 °C 12 h_v_2 was observed. In all measurements, several Raman spectra were taken to check that the γ-Fe_2O_3 was not transformed into α-Fe_2O_3 under laser heating.

For both attempts, although the morphology of the fabricated NPs is different, the composition of the iron oxide NPs is not so much. However, the proportion of each iron oxide phase in each sample is different depending on the attempt, observing a larger amount of γ-Fe_2O_3 than α-Fe_2O_3 in the sample prepared at the second attempt than at the first one.

- **Analysis by Raman spectroscopy of FeO$_x$ nanostructures**

By Raman spectroscopy, a more detailed analysis of the Raman signal for α-Fe_2O_3 and γ-Fe_2O_3 NPs can be carried out, similar to that of samples fabricated in air (see Sect. 4.3.3). For that, both the α-Fe_2O_3 and γ-Fe_2O_3 detected in the FeO$_x$ zones of samples Fe 10 Au 10 1100 °C 12 h_v_1 and Fe 10 Au 10 1100 °C 12 h_v_2, obtained at the first and second attempts, were studied. The Au/FeO$_x$ zones were not analyzed since for the sample Fe 10 Au 10 1100 °C 12 h_v_1 a high Au fluorescence did not allow identifying clearly the iron oxide spectra.

Figure 4.29 shows the α-Fe_2O_3 and γ-Fe_2O_3 average Raman spectra for the FeO$_x$ zones of both samples. It is possible to distinguish the seven phonon modes ($2A_{1g} + 5E_g$) allowed in Raman for α-Fe_2O_3, besides other bands [33–42]. An assignment of the Raman bands was achieved in Sect. 4.3.3. For γ-Fe_2O_3, besides the five phonon modes ($A_{1g} + E_g + 3T_{2g}$) allowed in Raman, other peaks related to the breaking of symmetry due to deviations at the stoichiometry are observed [49, 50]. The distinguished bands coincide with those measured by Ohtsuka et al. [51]. Despite identifying a large number of bands, according to the literature, it is only possible to assign three broad bands: around 365 cm^{-1} (T_{2g}), 510 cm^{-1} (E_g) and at 720 cm^{-1} (A_{1g}) [33, 43, 48, 52]. Table 4.7 presents the wavenumbers of Raman modes obtained from a lorentzian fitting and assigned according to the literature [33, 43, 48, 52].

Comparing Raman spectra, variations in the relative intensity, the wavenumber and at the FWHM of Raman bands can be observed. For α-Fe_2O_3 structures obtained at the first attempt (see Fig. 4.29a), the LO mode is clearly more intense with respect to the rest of bands than for α-Fe_2O_3 structures fabricated at the second attempt. As already indicated in Sect. 4.3.3, the LO mode is related to the structural disorder in the α-Fe_2O_3 NPs [34–36], thus it can be attributed to a decrease in the crystallinity and an increase of the concentration of defects, such as oxygen vacancies [53], for hematite structures obtained at the first attempt [38, 41]. Besides, for both α-Fe_2O_3

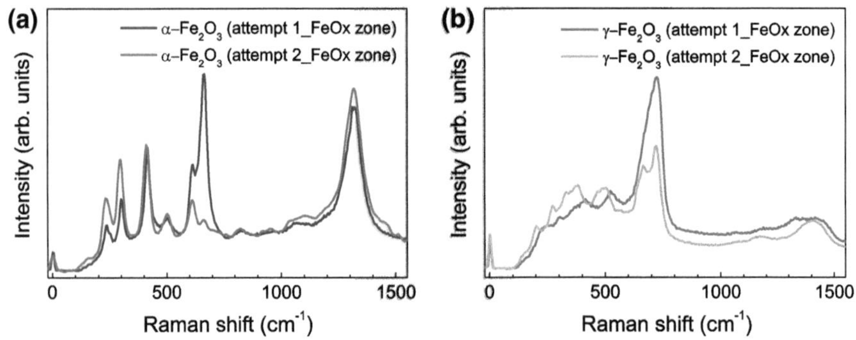

Fig. 4.29 **a** α-Fe_2O_3 and **b** γ-Fe_2O_3 average Raman spectra for FeO$_x$ zones of samples Fe 10 Au 10 1100 °C 12 h_v_1 and Fe 10 Au 10 1100 °C 12 h_v_2 obtained at the first and second attempt, respectively

Table 4.7 Raman bands and wavenumbers for α-Fe_2O_3 and γ-Fe_2O_3 of FeO$_x$ zones of samples Fe 10 Au 10 1100 °C 12 h_v_1 and Fe 10 Au 10 1100 °C 12 h_v_2

Raman bands α-Fe_2O_3	First attempt ν (cm^{-1})	Second attempt ν (cm^{-1})	Raman bands γ-Fe_2O_3	First attempt ν (cm^{-1})	Second attempt ν (cm^{-1})
$A_{1g}(1)$	235.0	227.2		237.6	202.4
$E_g(1)$	250.2	243.1		304.5	274.7
$E_g(2)$	294.4	292.5	T_{2g}	365.0	331.2
$E_g(3)$	308.1	303.2		412.1	382.5
$E_g(4)$	416.3	411.4		509.1	470.2
$A_{1g}(2)$	500.2	502.3	E_g	523.7	498.8
$E_g(5)$	613.9	610.9		674.9	667.7
Disorder	664.7	671.3	A_{1g}	728.2	721.2
Magnon	826.2	826.0		1187.1	1163.2
	980.1	957.1		1430.6	1403.4
	1067.3	1058.1			
	1106.1	1107.4			
Overtone	1322.9	1316.7			

and γ-Fe_2O_3 obtained at the first attempt, Raman bands are shifted towards larger wavenumbers than those of α-Fe_2O_3 and γ-Fe_2O_3 prepared at the second attempt. This confirms a decrease in the size of α-Fe_2O_3 and γ-Fe_2O_3 nanostructures [43] for sample obtained at the second attempt (see Fig. 4.29). These Raman results agree with the AFM results (see Sect. 4.4.2), confirming that structures obtained at the second attempt show are smaller than those obtained at the first attempt.

For these samples obtained in low vacuum, similar to nanostructures obtained in air (see Sect. 4.3.3), changes in the relative intensities of Raman bands are observed in α-Fe_2O_3 NPs, showing the different orientations. Figure 4.30a shows the in-plane Raman intensity image ($16 \times 16\ \mu m^2$) for the FeO$_x$ zone of the sample Fe 10 Au 10 1100 °C 12 h_v_1 obtained at the first attempt (see Fig. 4.25). The image was obtained from mapping the different single Raman spectra measured each 200 nm and integrating in the spectral range from 700 to 760 cm^{-1} to separate the signal corresponding to γ-Fe_2O_3 (blue color) and from 260 to 325 cm^{-1} (orange color) and 370 to 445 cm^{-1} (green color) to distinguish α-Fe_2O_3 structures with different orientation. The spectral range to discriminate the hematite signal with different orientation was chosen since it shows different relative intensities throughout the mapped area on sample. Single Raman spectra, shown in Fig. 4.30b, were taken from three points marked as 1, 2 and 3 on the Raman image that correspond to hematite NPs with different crystalline orientation.

Raman measurements were performed using the laser (532 nm) in p-polarization, as we explained in Chap. 2. Changing the polarization of the laser excitation on the three points marked as 1, 2 and 3 on the Raman image, we can observe variations in the relative intensities of Raman modes depending on the orientation of crystal

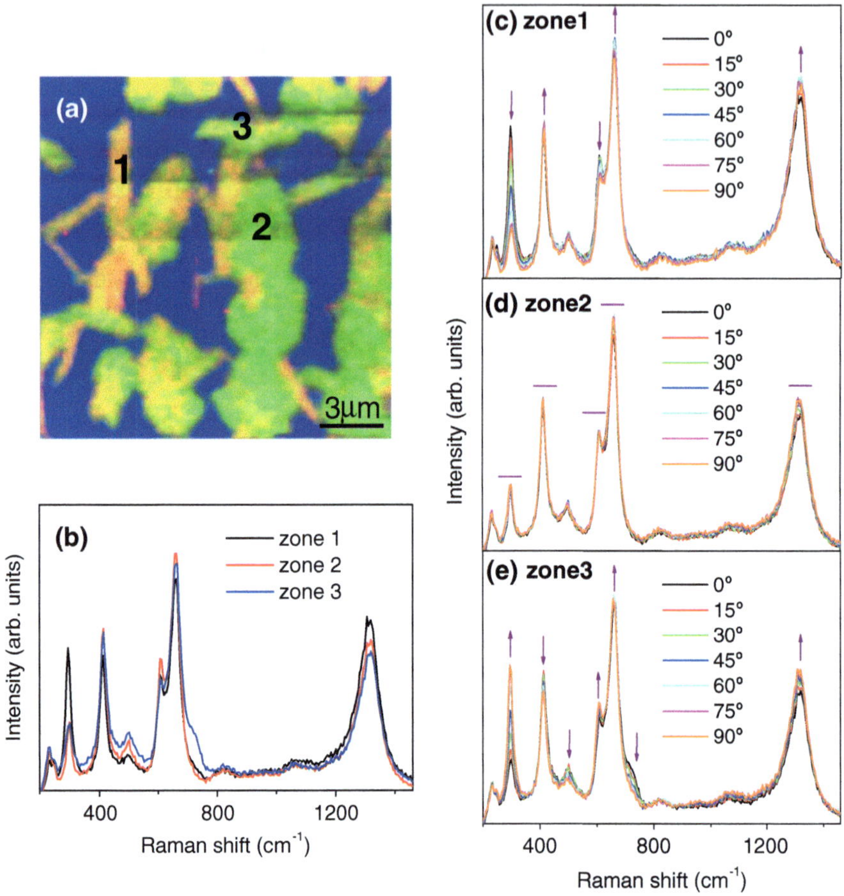

Fig. 4.30 **a** In-plane Raman intensity image for the FeO$_x$ zone of the sample Fe 10 Au 10 1100 °C 12 h_v_1 obtained from mapping the different single Raman spectra measured each 200 nm, **b** Raman spectra on three points marked as 1, 2 and 3 on the Raman image that correspond to hematite NPs with different crystalline orientation and **c**, **d** and **e** single Raman spectra on the three points changing the polarization of the laser excitation

planes. Figure 4.30c, e shows that the Raman spectra measured on zone 1 and 3, which correspond to the microwires, are sensitive to the orientation of crystal planes. However modifications in Raman spectrum on zone 2 are not observed (Fig. 4.30d), thus this region is not sensitive to changes in the laser polarization due to the spacial orientation of α-Fe_2O_3 crystals.

4.4.4 Optical Characterization of Samples Prepared in Low Vacuum

Figure 4.31 shows the optical absorption spectra of samples Fe 10 Au 10 1100 °C 12 h_v_1 and Fe 10 Au 10 1100 °C 12 h_v_2, referenced to a silica substrate in order to obtain the net optical absorption of Au, FeO_x and Au/FeO_x zones. For the sample obtained at the first attempt (Fig. 4.31a), it is possible to analyze the three zones corresponding to Au, FeO_x and Au/FeO_x NPs. The absorption spectrum for the Au NPs shows a shoulder in the range 400–500 nm that is due to Au interband transitions, promoting electrons from the 3d band to the Fermi level [54]. Besides, the absorption band corresponding to the SPs, around 553 nm, is clearly observed. For FeO_x zone, the absorption bands related to electronic transitions can be distinguished and identified according to the literature [21, 46]. An assignment of these bands was already carried out in Sect. 4.3.4. Briefly, the absorption bands in the region between 200 and 400 nm results from ligand to metal charge transfer transitions and partly from the contributions of Fe^{3+} ligand field transitions, and in the region between 400 and 600 nm, the absorption bands correspond to the pair excitation processes and contributions of ligand field transitions. Finally, for Au/FeO_x zone, it is possible to differentiate the absorption bands corresponding to the iron oxide and to the SPs around 571 nm. This band is shifted towards larger wavelengths and is wider than the SP band obtained for the single Au NPs of the Au zone. These modifications of the position of SP band are due to the variations in the surrounding medium of the Au NPs. In the Au zone, the Au NPs are surrounded by air while in the Au/FeO_x zone, they are surrounded by air and FeO_x NPs. Since dielectric constant of the iron oxide is higher than that of the air [44], the SP band is shifted towards larger wavelengths [20]. The morphological features of Au NPs can also affect to the SP position. On the other hand, the SP band of Au NPs in the Au/FeO_x zone can be decomposed into two gaussian bands corresponding to a SP band around 554 nm

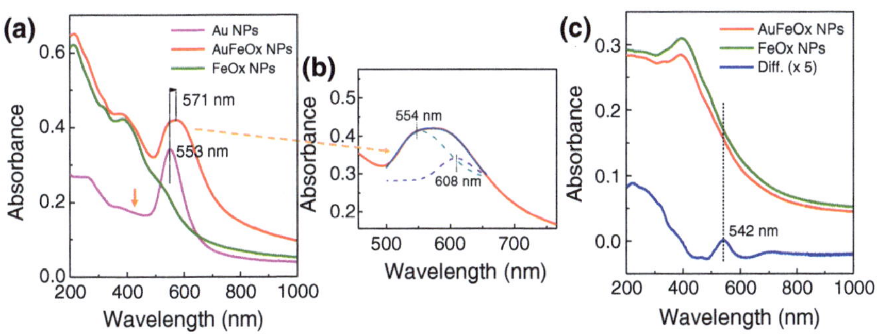

Fig. 4.31 Optical absorption spectra for the different zones of samples obtained at the **a** first attempt and **c** second attempt, from a 10 nm Au film and a 10 nm Fe film annealed at 1100 °C for 12 h in low vacuum. **b** The fit (*blue line*) to two gaussian bands of the SP absorption band (*red line*) for Au/FeO_x zone obtained at the first attempt

associated with the Au NPs surrounded by air (SPs in the Au zone are observed at 553 nm) and the other one related to Au NPs surrounded by FeO_x NPs at 608 nm, as shown in Fig. 4.31b. Thus, SP band is widened since it is composed of two SP bands associated with two different media. A high size dispersion of Au NPs or small Au NPs in the Au/FeO_x zone can also widen the SP band. With respect to the intensity of SP bands, we observe that the SP band in the Au zone is more intense (~0.20) than in the Au/FeO_x zone (~0.15). This can be related to smaller Au NPs in the Au/FeO_x zone of sample.

At the second attempt (see Fig. 4.31c), we did not obtain Au NPs in the Au zone (see Fig. 4.18), thus in this case we analyze the optical absorptions of the two zones corresponding to FeO_x and Au/FeO_x NPs. For both regions, similar spectra are observed, distinguishing the transitions bands corresponding to the iron oxide NPs. Normalizing both curves to the maximum (at 396 nm) and subtracting the signal of FeO_x zone to that of the Au/FeO_x zone, an absorption band is obtained around 542 nm, which can be associated with the SP absorption band of Au NPs. This weak SP band was hidden by transitions bands of the iron oxide NPs due to the small amount of Au NPs present in the Au/FeO_x zone. This result confirms the Raman measurements (Sect. 4.4.3), where we observed Au NPs but their fluorescence signal was lower than in the Au/FeO_x zone of the sample fabricated at the first attempt.

4.5 Conclusions

In summary, we have presented here a new method to fabricate large amounts of complex nanoparticles covering large (several centimeters) substrates by thin film deposition and annealing in air and low vacuum. Within this method, different strategies were performed, which allow tuning the morphology of the nanoparticles independently or promoting the formation of complex dimers to enhance the coupling effects. In this way, surface plasmons can be modified by tuning the morphological properties of complex NPs. The method was successfully applied to the fabrication of Au/α-Fe_2O_3 and Au/α-Fe_2O_3-γ-Fe_2O_3 nanostructures ensembles, although the method might be applied to a wide range of composed materials.

References

1. M.V. Uwe Kreibig, Optical properties of metal clusters, springer series in material science, vol. 25 (Springer, Berlin, 1995)
2. A. Alivisatos, Semiconductor clusters, nanocrystals, and quantum dots. Science **271**(5251), 933–937 (1996)
3. P. Tartaj, M. Morales, T. Gonzalez-Carreño, S. Veintemillas-Verdaguer, C. Serna, The iron oxides strike back: from biomedical applications to energy storage devices and photoelectrochemical water splitting. Adv. Mater. **23**(44), 5243–5249 (2011)

4. S. Chatterjee, A. Bandyopadhyay, K. Sarkar, Effect of iron oxide and gold nanoparticles on bacterial growth leading towards biological application. J. Nanobiotechnol. **9**, (2011)
5. Y. Lee, M. Garcia, N. Frey Huls, S. Sun, Synthetic tuning of the catalytic properties of au-fe3o 4 nanoparticles. Angew. Chem. Int. Ed. **49**(7), 1271–1274 (2010)
6. M. Grzelczak, J. Pérez-Juste, P. Mulvaney, L. Liz-Marzá, Shape control in gold nanoparticle synthesis. Chem. Soc. Rev. **37**(9), 1783–1791 (2008)
7. H.L. Liu, J.H. Wu, J.H. Min, Y.K. Kim, Synthesis of monosized magnetic-optical aufe alloy nanoparticles. J. Appl. Phys. **103**, 707D529 (2008)
8. H.-Y. Park, M.J. Schadt, L. Wang, I.-I.S. Lim, P.N. Njoki, S.H. Kim, M.-Y. Jang, J. Luo, C.-J. Zhong, Fabrication of magnetic core@shell Fe oxide@Au nanoparticles for interfacial bioactivity and bio-separation. Langmuir: ACS J. Surf. Colloids **23**(17), 9050–9056 (2007)
9. R.S. Kane, Fabricating complex polymeric micro- and nanostructures: lithography in microfluidic devices. Angew. Chem. (Int. Ed. Engl.) **47**(8), 1368–1370 (2008)
10. A.E.B. Presland, G.L. Price, D.L. Trimm, Hillock formation by surface diffusion on thin silver films. Surf. Sci. **29**(2), 424–434 (1972)
11. M.S. Rahman Khan, Hillock and island formation during annealing of gold films. Bull. Mater. Sci. **9**, 55–60 (1987)
12. M.S. Jackson, C.-Y. Li, Stress relaxation and hillock growth in thin films. Acta Metall. **30**(11), 1993–2000 (1982)
13. C.V. Thompson, Solid-State dewetting of thin films. Ann. Rev. Mater. Res. **42**(1), 399–434 (2012)
14. J.A. Thornton, D.W. Hoffman, Stress-related effects in thin films. Thin Solid Films **171**(1), 5–31 (1989)
15. I. Doron-Mor, Z. Barkay, N. Filip-Granit, A. Vaskevich, I. Rubinstein, Ultrathin gold island films on silanized glass. morphology and optical properties. Chem. Mater. **16**(18), 3476–3483 (2004)
16. S. Pillai, K.R. Catchpole, T. Trupke, M.A. Green, Surface plasmon enhanced silicon solar cells. J. Appl. Phys. **101**(9), 093105 (2007)
17. S. Aggarwal, A. Monga, S. Perusse, R. Ramesh, V. Ballarotto, E. Williams, B. Chalamala, Y. Wei, R. Reuss, Spontaneous ordering of oxide nanostructures. Science **287**(5461), 2235–2237 (2000)
18. S. Aggarwal, S. Ogale, C. Ganpule, S. Shinde, V. Novikov, A. Monga, M. Burr, R. Ramesh, V. Ballarotto, E. Williams, Oxide nanostructures through self-assembly. Appl. Phys. Lett. **78**(10), 1442–1444 (2001)
19. S. Shinde, A. Ogale, S. Ogale, S. Aggarwal, V. Novikov, E. Williams, R. Ramesh, Self-organized pattern formation in the oxidation of supported iron thin films. I. An experimental study. Phys. Rev. B **64**(3), 1–5 (2001)
20. M.A. Garcia, Surface plasmons in metallic nanoparticles: fundamentals and applications. J. Phys. D: Appl. Phys. **44**(28), 283001 (2011)
21. R.M. Cornell, U. Schwertmann, *The Iron Oxides: Structure, Properties, Reactions, Occurrence and Uses*, 2nd edn. (WILEY-VCH Verlag GmbH & Co. KGaA, Weinheim, 2003)
22. W.H. Hung, M. Aykol, D. Valley, W. Hou, S.B. Cronin, Plasmon resonant enhancement of carbon monoxide catalysis. Nano Lett. **10**(4), 1314–1318 (2010)
23. C. de Julián Fernández, G. Mattei, E. Paz, R. Novak, L. Cavigli, L. Bogani, F. Palomares, P. Mazzoldi, A. Caneschi, Coupling between magnetic and optical properties of stable au-fe solid solution nanoparticles. Nanotechnology **21**, 16 (2010)
24. Y.S. Touloukian, R.K. Kirby, R.E. Taylor, P.D. Desai, Thermal Expansion: Metallic Elements and Alloys (Thermophysical Properties of Matter, vol. 12). IFI/Plenum. 1 edition, 1975
25. E.C. Romani, D. Vitoreti, P.M.P. Gouvêa, P.G. Caldas, R. Prioli, S. Paciornik, M. Fokine, A.M.B. Braga, A.S.L. Gomes, I.C.S. Carvalho, Gold nanoparticles on the surface of soda-lime glass: morphological, linear and nonlinear optical characterization. Opt. Express **20**(5), 5429–5439 (2012)
26. H. Nelson, The low temperature oxidation of iron. J. Chem. Phys. **6**(10), 606–611 (1938)

27. F. Kenfack, H. Langbein, Influence of the temperature and the oxygen partial pressure on the phase formation in the system Cu–Fe–O. Cryst. Res. Technol. **39**(12), 1070–1079 (2004)
28. G. Rhead, Surface self-diffusion of silver in various atmospheres. Acta Metall. **13**(3), 223–226 (1965)
29. S.K. Sharma, J. Spitz, Hillock formation, hole growth and agglomeration in thin silver films. Thin Solid Films **65**(3), 339–350 (1980)
30. B.E.H. Oswald Kubaschewski, Oxidation of metals and alloys (Butterworths, London, 1967)
31. See http://www.jobinyvon.com/usadivisions/Raman/applications/Fluorescence01.pdf
32. T. Hasegawa, J. Nishijo, J. Umemura, Separation of raman spectra from fluorescence emission background by principal component analysis. Chem. Phys. Lett. **317**(6), 642–646 (2000)
33. D.L.A. de Faria, S. Venancio Silva, M.T. De Oliveira, Raman microspectroscopy of some iron oxides and oxyhydroxides. J. Raman Spectrosc. **28**(11), 873–878 (1997)
34. K. McCarty, Inelastic light scattering in α-Fe_2O_3: phonon vs magnon scattering. Solid State Commun. **68**(8), 799–802 (1988)
35. H. Cao, G. Wang, L. Zhang, Y. Liang, S. Zhang, X. Zhang, Shape and magnetic properties of single-crystalline hematite (alpha-Fe_2O_3) nanocrystals. Chemphyschem: Eur. J. Chem. Phys. Phys. Chem. **7**(9), 1897–1901 (2006)
36. I. Cesar, K. Sivula, A. Kay, R. Zboril, M. Gratzel, Influence of feature size, film thickness, and silicon doping on the performance of nanostructured hematite photoanodes for solar water splitting. J. Phys. Chem. C **113**(2), 772–782 (2009)
37. T. Martin, R. Merlin, D. Huffman, M. Cardona, Resonant two magnon Raman scattering in α-Fe_2O_3. Solid State Commun. **22**(9), 565–567 (1977)
38. D. Bersani, P. Lottici, A. Montenero, Micro-raman investigation of iron oxide films and powders produced by sol-gel syntheses. J. Raman Spectrosc. **30**(5), 355–360 (1999). cited By (since 1996)154
39. F.J. Owens, J. Orosz, Effect of nanosizing on lattice and magnon modes of hematite. Solid State Commun. **138**(2), 95–98 (2006)
40. M. Lübbe, A.M. Gigler, R.W. Stark, W. Moritz, Identification of iron oxide phases in thin films grown on Al_2O_3(0001) by Raman spectroscopy and X-ray diffraction. Surf. Sci. **604**(7–8), 679–685 (2010)
41. I.V. Chernyshova, M.F. Hochella Jr, A.S. Madden, Size-dependent structural transformations of hematite nanoparticles. 1. Phase transition. Phys. Chem. Chem. Phys. **9**(14), 1736–1750 (2007)
42. M. Legodi, D. Dewaal, The preparation of magnetite, goethite, hematite and maghemite of pigment quality from mill scale iron waste. Dyes Pigm. **74**(1), 161–168 (2007)
43. A.M. Jubb, H.C. Allen, Vibrational spectroscopic characterization of hematite, maghemite, and magnetite thin films produced by vapor deposition. ACS Appl. Mater. Interfaces **2**(10), 2804–2812 (2010)
44. See http://www.flowmeterdirectory.com/dielectricconstant01.html
45. M.A. García, J. Llopis, S.E. Paje, A simple model for evaluating the optical absorption spectrum from small au-colloids in sol-gel films. Chem. Phys. Lett. **315**(5–6), 313–320 (1999)
46. Y. He, Y. Miao, C. Li, S. Wang, L. Cao, S. Xie, G. Yang, B. Zou, C. Burda, Size and structure effect on optical transitions of iron oxide nanocrystals. Phys. Rev. B—Condens. Matter Mater. Phys. 71(12), (2005)
47. C. Yan, Y. Chen, A. Jin, M. Wang, X. Kong, X. Zhang, Y. Ju, L. Han, Molecule oxygen-driven shaping of gold islands under thermal annealing. Appl. Surf. Sci. **258**(1), 377–381 (2011)
48. M. Hanesch, Raman spectroscopy of iron oxides and (oxy)hydroxides at low laser power and possible applications in environmental magnetic studies. Geophys. J. Int. **177**(3), 941–948 (2009)
49. N. Boucherit, A.H.-L. Goff, S. Joiret, Raman studies of corrosion films grown on Fe and Fe-6Mo in pitting conditions. Corros. Sci. **32**, 497–507 (1991)
50. A. Demoulin, C. Trigance, D. Neff, E. Foy, P. Dillmann, V. L'Hostisd, The evolution of the corrosion of iron in hydraulic binders analysed from 46- and 260-year-old buildings. Corros. Sci. **52**(10), 3168–3179 (2010)

51. T. Ohtsuka, K. Kubo, N. Sato, Raman spectroscopy of thin corrosion films on iron at 100 to 150°C in air. Corrosion-NACE **42**(8), 476–481 (1986)
52. I. Chamritski, G. Burns, Infrared- and Raman-active phonons of magnetite, maghemite, and hematite: a computer simulation and spectroscopic study. J. Phys. Chem. B **109**(11), 4965–4968 (2005)
53. N.D. Phu, D.T. Ngo, L.H. Hoang, N.H. Luong, N. Chau, N.H. Hai, Crystallization process and magnetic properties of amorphous iron oxide nanoparticles. J. Phys. D: Appl. Phys. **44**(34), 345002 (2011)
54. A. Mooradian, Photoluminescence of metals. Phys. Rev. Lett. **22**, 185–187 (1969)

Chapter 5
Combination of Surface Plasmon Resonance and X-Ray Absorption Spectroscopies. Experimental Setup

This chapter presents the development of a new setup to combine surface plasmon resonance and X-ray absorption spectroscopy. The system allows the study of the interaction between electromagnetic radiation and matter using one type of radiation to modify the material and the other one as a probe, performing the study in real time and in situ. The design details and the type of experiments that can be carried out are showed in this chapter. The device has been mounted at the SpLine BM25 beamline at ESRF in Grenoble, France, and it is currently available for experiments.

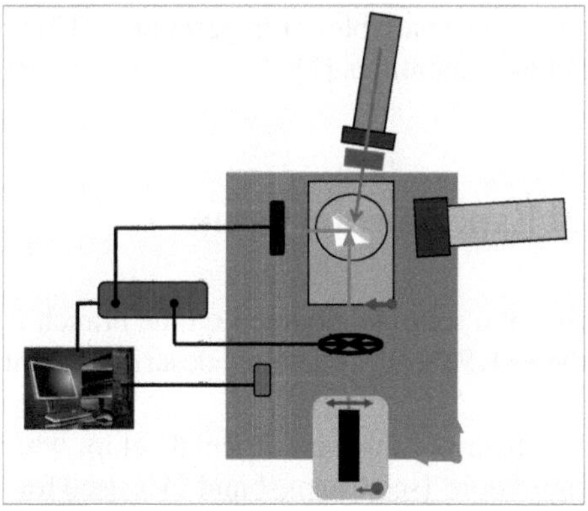

5.1 Introduction

As already indicated in Chap. 1, surface plasmon resonance (SPR) in metallic thin films is extremely sensitive to modifications in the properties and/or morphology of the metallic film, the substrate and the top dielectric medium. Thus, SPR spectroscopy is considered an optical sensor by means of an evanescent field [1] and is commonly used as probe to study *in situ* the growth or modifications

A. Serrano Rubio, *Modified Au-Based Nanomaterials Studied by Surface Plasmon Resonance Spectroscopy*, Springer Theses, DOI 10.1007/978-3-319-19402-8_5

of dielectric materials [2]. Similarly, X-ray absorption spectroscopy (XAS) is an invaluable technique that provides detailed information on the electronic structure of materials [3, 4]. It is also known that X-rays induce modifications in many kinds of materials [5, 6]. Therefore, the combination of both techniques may be used to study the interaction between electromagnetic radiation and matter by using one of the beams to modify the material and the other one as a probe, performing the study in real time and *in situ*. Despite the very different energy ranges of SPR and XAS (∼eV for SPR and ∼keV for XAS), both processes deal with the excitation of electrons to about the Fermi level. Since SPR alters the electronic population of the metallic film around the Fermi level, a modification of the XAS spectra could be expected. Similarly, when electrons are pumped to above the Fermi level by X-ray absorption, one might expect variations in the SPR spectra. Given the sensitivity of SPR, it can be also used to study the effects induced by X-rays on a large variety of materials such as glasses, crystalline oxides or organic matter [7].

The combination of both techniques, to our best knowledge, has never been achieved and is technologically challenging. Optical excitation of SPR can be performed only with very special geometries (see Sect. 1.3.2.4 in Chap. 1), hindering its combination with other techniques. Here, we show a SPR system compatible with a XAS beamline (SPR-XAS setup). With this setup we have explored the combination of both techniques and tried to determine the types of experiments that can be carried out and the information we can extract from them. The device has been mounted at the SpLine BM25 beamline at European Synchrotron Radiation Facility (ESRF) in Grenoble, France, and is now available for experiments. This chapter describes the experimental setup and its capabilities [7].

5.2 Description of Experimental Setup

The SPR-XAS experimental setup is mounted on the branch A of the CRG BM25-SpLine Beamline at the ESRF [8, 9]. A detailed description of the Branch A features is shown in Chap. 2.

Figure 5.1 shows a scheme of the experimental setup, where the Kretschmann-Raether [10–12] configuration[1] (see Chaps. 1 and 2) is used for the optical excitation of SPR. Briefly, for this configuration, a metallic film is grown on a glass substrate that is coupled to a prism on the glass side (see inset in Fig. 5.1). The glass/metal interface is illuminated with a laser beam in total reflection conditions and an evanescent field propagates through the metallic layer reaching the other side where surface plasmons (SPs) are excited. As already indicated in Chap. 2, this configuration limits considerably the thickness of the metallic film to a narrow window ∼50 ± 10 nm [2, 10].

[1]The Otto configuration [10, 13] is also widely employed for the excitation of SPR. However, it requires that the metal surface and the prism are separated a certain distance. Therefore, it is almost impossible to illuminate with X-rays the same region where the SPR is excited.

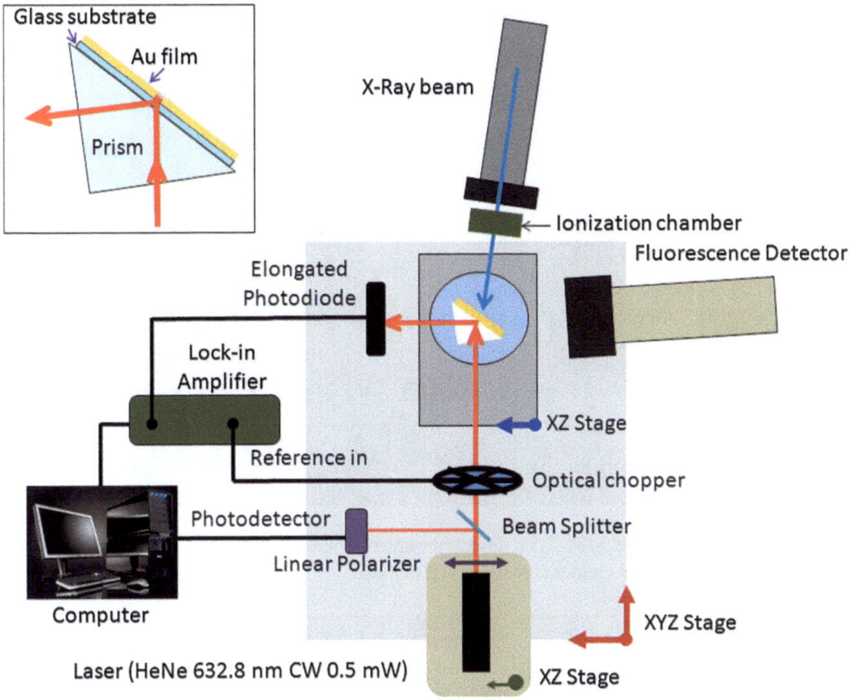

Fig. 5.1 Scheme of the SPR-XAS device for simultaneous SPR and XAS measurements. The *inset* shows a detail of the illumination of the glass/metal interface

Figure 5.2 displays a photograph of the SPR-XAS device mounted in the experimental hutch, showing the different elements of the experimental setup. Figure 5.3 shows more pictures of the system viewed from different angles.

In this device [7], the SPR system is mounted onto the optical table of the XAS system. The translation and rotation motors enable an accurate positioning of the whole SPR device system with respect to the X-ray beam. Excitation is performed with a HeNe (632.8 nm) linearly polarized laser (power stabilized, CW 0.5 mW) in a p-polarization geometry with respect to the sample surface. The laser spot size is 0.3 mm, measured as the full width half maximum (FWHM) of the Gaussian profile of the spot. The laser is mounted on an independent XZ translation stage and on a cradle equipped with yaw and pitch movements for an accurate orientation of the beam on the sample. A beam splitter deflects about 5 % of the laser intensity to a photodetector in order to record fluctuations in laser intensity during the experiments. These fluctuations are less than 0.05 % over several hours. Beyond the beam splitter, the laser beam is modulated with an optical chopper working at 479 Hz. The sample, consisting of a thin metallic film (in our case Au thin films) grown on a glass substrate plus possible overlayers of dielectric materials, is fixed to a triangular quartz prism using a gel index matching for a good coupling. Sample and prism are mounted on

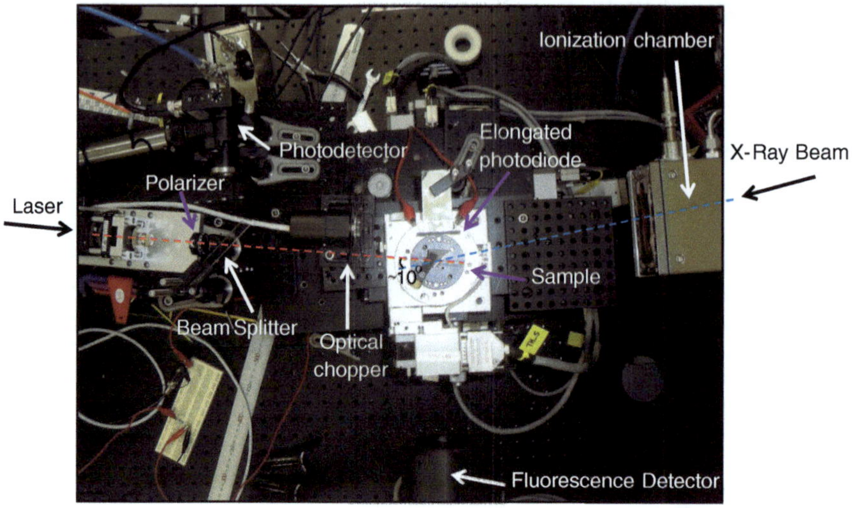

Fig. 5.2 Photograph of the SPR-XAS device mounted in the experimental hutch at the SpLine beamline, with the different elements marked and named

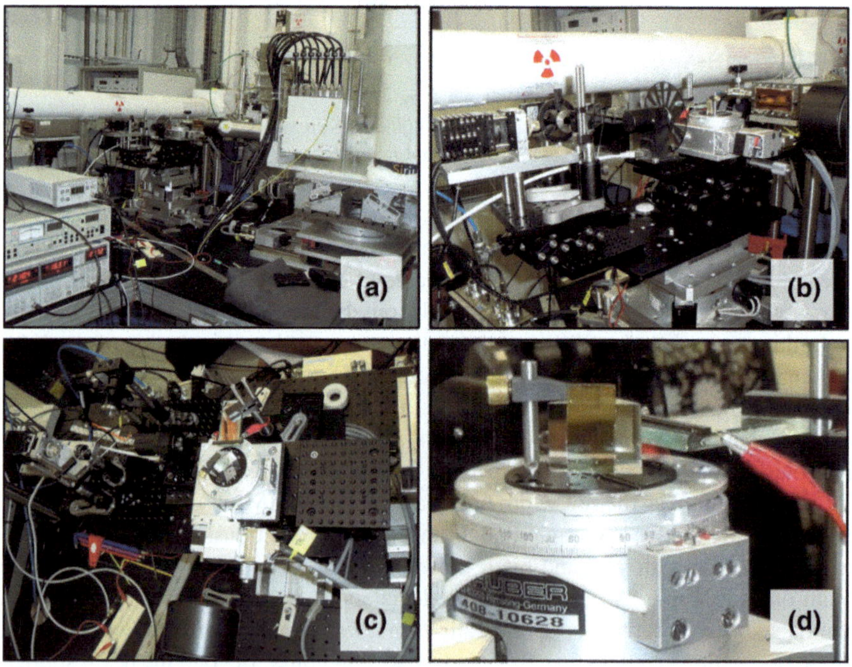

Fig. 5.3 Photographs of the SPR-XAS experimental setup viewed from different angles

top of a rotating motor that allows varying the laser incidence angle (see Figs. 5.2 and 5.3d). This sample stage has an independent XZ translation stage for sample positioning. During the measurements, the laser beam reflected at the sample is collected as a function of the incidence angle with an elongated photodiode (to avoid moving the detector during the angle scan). The photodiode signal is registered with a lock-in amplifier using the frequency of the optical chopper as reference. All the elements that comprise the SPR device (motors, laser monitor and lock-in amplifier signals) are integrated and synchronized in the software for the control of the XAS system at the SpLine beamline. To avoid the X-ray passing through the optical prism and reach the laser cavity, the laser and X-ray beams are slightly tilted ($\sim 10°$), as shown in Figs. 5.1 and 5.2. Note that, for example, for 14 keV X-rays about 0.1 % of the radiation is transmitted through 5 mm of silica. This small fraction may have significant effects in the instrumentation due to the high intensity of the X-ray beam. Actually, it is observed that without tilting both beams, the X-ray photons reaching the laser induce thermal fluctuations leading to variations in laser intensity of the order of 0.5 %, despite its power is stabilized.

On the other hand, the sample can be electrically isolated or grounded to avoid its electrical charging, caused by the photoemission of electrons upon X-ray irradiation. However, within the resolution of this experimental setup, no difference is found between the spectra obtained with the sample grounded or not.

All the elements employed in this SPR system, except to the laser, are the same employed in the experimental setup mounted in the Departamento de Electrocerámica at the Instituto de Cerámica y Vidrio (CSIC) of Madrid (see Sect. 2.10 in Chap. 2).

5.2.1 Alignment of the Beams

As explained in Sect. 2.10 of Chap. 2, SPR spectroscopy in the Kretschmann-Raether configuration is performed by collecting the reflected light as a function of the incident angle while rotating the sample [10]. In this configuration, just the points along the rotation axis remain in the same position while the sample rotates (see Fig. 5.4). Additional difficulties arise when aligning properly both the laser and X-ray beams. If the laser spot does not impinge at the sample at this rotation axis, its position will change during the angle scan. This is not critical in standard SPR experiments when samples are homogeneous. However, for this particular device, if the laser and X-ray beams do not impinge both at the same position of the rotation axis, they will become misaligned as the sample rotates, illuminating different points at the sample surface, as illustrated in Fig. 5.4. Thus, it is crucial to ensure that the laser spot illuminates the sample surface at the rotation axis. This can be confirmed in a simple way using two lasers. If they are properly aligned at the rotation axis, no misalignment will take place when the sample is rotated.

Once the laser is properly aligned at the film surface, the X-ray beam must be aligned at the same position. For this purpose, the whole SPR system is moved with respect to the X-ray beam using the XAS system translation stage. The alignment

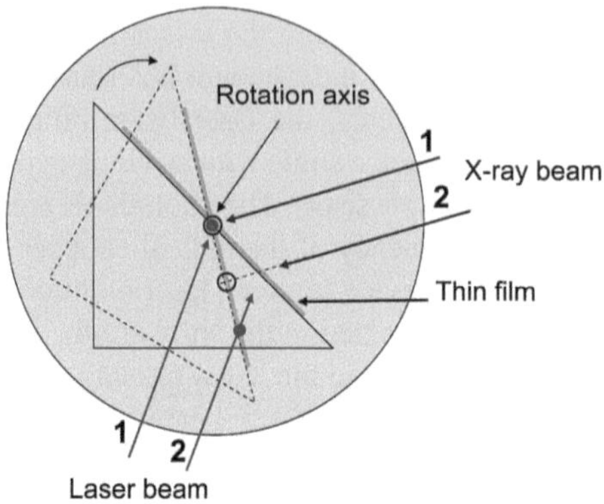

Fig. 5.4 Illustration of the possible misalignment of the laser and X-ray beams. After the rotation of the prism, X-ray and laser beams illuminate the surface of the prism at the *blue empty circle* and at the *red filled circle*, respectively. If both beams are initially coincident at the rotation axis (case 1), they will remain aligned after the rotation of the prism (*dashed prism*). This is the desired situation. If the beams are initially aligned out of the rotation axis (case 2), they will intersect the prism surface at different points after the rotation of the prism

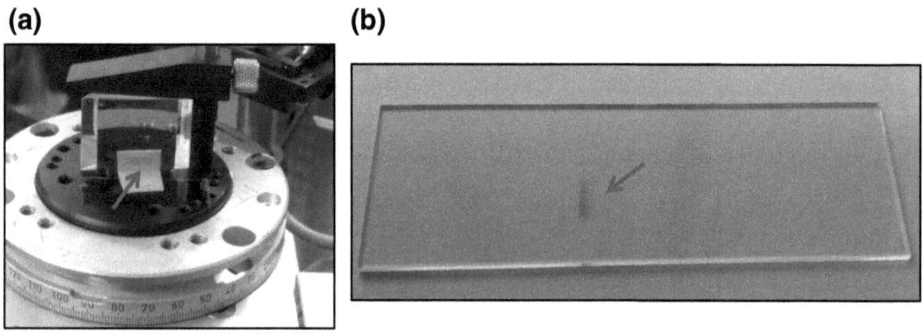

Fig. 5.5 Photographs of the darkening effects induced by X-ray irradiation in **a** a photosensitive paper and **b** a soda-lime glass

can be performed by using a thin phosphor film or screen and a camera viewing the sample to observe simultaneously the X-rays and laser spots. A toggle-shutter for the incident laser beam may be useful if its light overwhelms the X-rays spot on the phosphor. Alternatively, a material sensitive to X-rays such as a photosensitive paper can be placed at the sample position and checked after irradiation if the laser beam spot matches the spot induced by the X-rays. The alignment can be also carried out taking advantage of the fact that hard X-rays induce darkening of some glasses [5], as Fig. 5.5 illustrates. After irradiating for few minutes, the glass exhibits a dark spot that can be also used to check the proper alignment with the laser. In our experiments, to perform the alignment of the beams, photosensitive paper was used for samples with silica substrates and soda-lime glasses for samples with this kind of substrates.

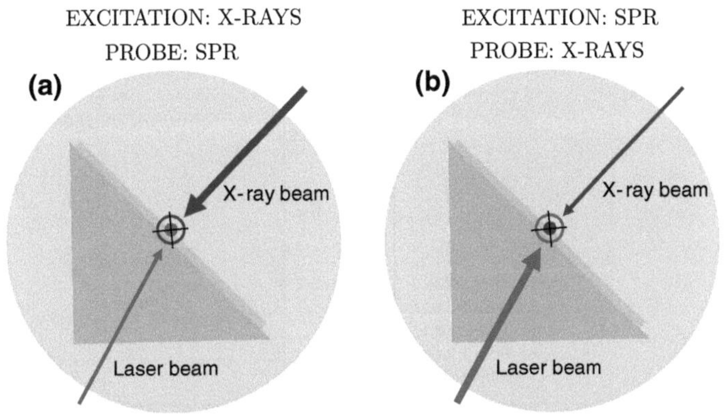

Fig. 5.6 Scheme of X-ray and laser beams illuminating the sample. **a** X-ray beam spot is larger than that of the laser beam and **b** laser beam spot is larger than that of the X-ray beam

Once the laser and X-ray beams are aligned, they can reach different regions of the sample using the sample Z translation stage of the sample. Laser and X-ray beam will remain aligned at different points of the sample but, since the displacement is vertical, this new point will still be in the rotation axis. With this system, it is also possible to change from one sample to another without modifying the alignment.

5.2.2 Spot Size of the Beams

As explained above, some misalignment between the spots of the laser and the X-rays, when scanning the angle, is unavoidable for triangular prisms due to the refraction. This problem can be solved if it is ensured that the spot of the excitation beam (laser or X-ray beam) is larger than the spot of the probe (laser or X-ray beam). In this way, we improve the quality of the measurements and we are sure that we are probing an area which is being completely excited or modified. To study the effect of X-ray irradiation on the SPR curve, the X-ray spot must be larger than that of the laser, so that the SPR spectrum is collected from a region fully illuminated with X-rays. In a similar way, to study the effect of SPR excitation on the XAS spectrum, the spot of the X-ray beam must be smaller than the laser spot, to ensure that the recorded XAS spectrum corresponds to a region where SPR is excited (see Fig. 5.6).

5.3 Resolution of the Experimental Setup

The resolution of the SPR-XAS device depends on the type of measurement performed. We can distinguish between reflectivity and XAS measurements.

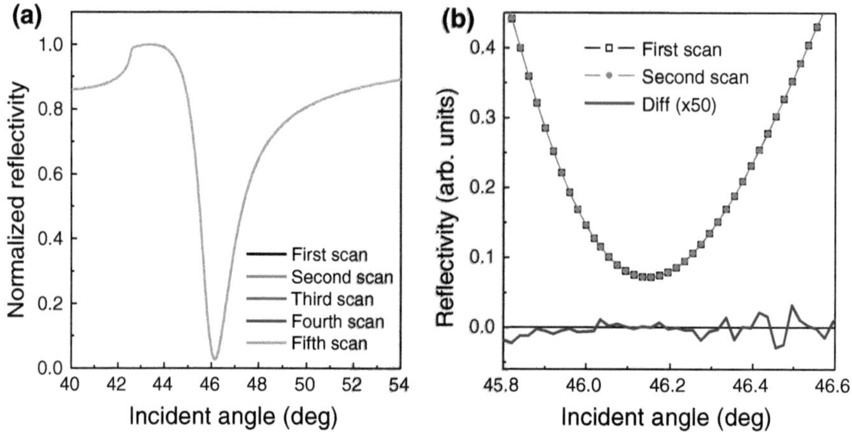

Fig. 5.7 **a** Consecutive spectra obtained with the SPR device shown in Fig. 5.1; the curves overlap and are not distinguished in the graph. **b** A detail of the resonance region between the first and second scans and the difference in the scans (multiplied ×50)

5.3.1 Reproducibility of SPR Spectra

Figure 5.7a shows five consecutive SPR spectra for a 45 nm Au film grown onto a glass substrate and collected in the same conditions without X-ray irradiation. The relative difference between the first and second spectra (see Fig. 5.7b) is of the order of $\Delta R/R = 10^{-3}$ for the region of the spectrum with highest slope and of the order of 10^{-4} for the rest of the spectrum[2]. Therefore, with this equipment we can clearly detect relative differences below 0.1 % with single scans. The resolution of the measurement can be improved upon accumulation of scans (scans in Fig. 5.7 take 30 min each one). Moreover, no drift was found in the resonance position for ten consecutive scans up to a resolution of 0.01°, which is the minimum shift that this setup could detect.

5.3.2 Reproducibility of XAS Spectra

Figure 5.8 shows two XAS spectra at the Au L3 edge relative to a 45 nm Au film grown onto a glass substrate, without SPR excitation. In this case, the difference between both scans is of the order of 10^{-2}, which is the noise level for the integration time used (30 s per point). The resolution of measurements can be improved changing

[2]For the SPR-XAS setup, it is possible to detect relative differences in the SPR spectra of at least one order of magnitude smaller than for the SPR setup developed at the Instituto de Cerámica y Vidrio of Madrid.

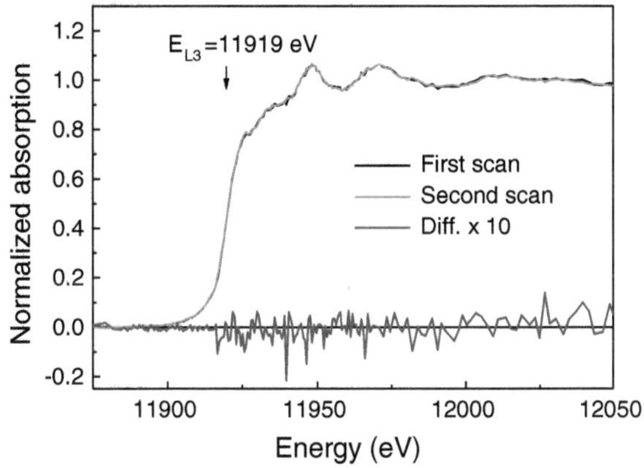

Fig. 5.8 Consecutive XAS spectra measured at the Au L3 edge on a 45 nm Au film without excitation of SPR and the difference between both scans

the integration time, accumulating more scans or using a laser with a spot size larger than 0.3 mm (spot size of laser used).

5.4 Capabilities of the Experimental Setup

Combining SPR and XAS allows us to perform different kinds of experiments, recording SPR or XAS signals while scanning other parameters such as time, energy, etc. Some examples are presented here, in order to illustrate the different configurations that can be used and the information that can be extracted from them. Some of those examples are studied in detail in the following chapters.

5.4.1 Study of SPR upon X-Ray Irradiation

Figure 5.9 shows the SPR spectra for Au thin films grown on different glass substrates measured before and after X-ray irradiation for 200 min using X-rays at E = 11.95 keV (above the Au L3 edge).

For soda-lime substrates (Fig. 5.9a), we can observe that X-ray irradiation induces a decrease in the reflected intensity for the whole SPR curve. The effect is accumulative upon several scans and is not related to SPR of Au film but to the darkening of the glass substrate upon X-ray irradiation [5]. As mentioned above, SPR is very sensitive to changes in the properties of the metallic film and the dielectric media close to the film [1, 2]. Therefore, modifications induced in the soda-lime substrates upon X-ray irradiation can be tracked by SPR spectroscopy, as we show in the following chapter. It is well known that hard X-ray irradiation induces color centers in soda-lime glasses [5, 14, 15]. These defects are commonly associated with network formers or modifiers (as Na or K in soda-lime and B in the borosilicates) that change

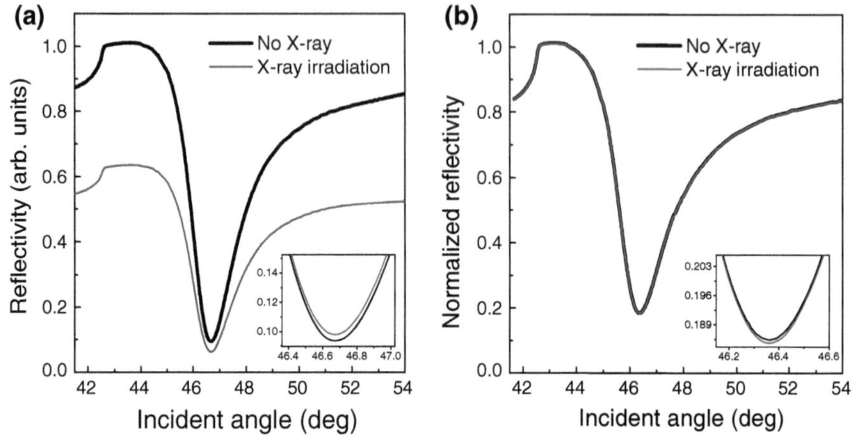

Fig. 5.9 SPR curves for a 45 nm Au film grown onto **a** soda-lime and **b** silica substrates before and after X-ray irradiation for 200 min using X-rays at E = 11.95 keV (above the Au L3 edge)

their coordination and oxidising state upon irradiation, leading to the formation of color centers responsible of the glass darkening [16]. Besides the darkening, X-rays induce a slight modification in the SPR spectra, increasing the reflectivity about 0.003 and narrowing the resonance around 0.04° at FWHM, as shown the inset of Fig. 5.9a. Contrary to soda-lime substrates, for silica substrates (Fig. 5.9b), no significant variations are found in the reflected intensity upon X-ray irradiation. However, a detail of the resonance region (see inset in Fig. 5.9b) reveals a small decrease of the reflectivity around 0.001 and a widening of the curve around 0.02° at FWHM. These modifications of SPR curves for both cases can be studied to determinate the changes in both real and imaginary part of the refractive index with great accuracy.

With this experimental setup, kinetics in real time and *in situ* can be also studied, for example the kinetics of formation and fading of color centers in the soda-lime substrates [7]. For that purpose, the incident angle is fixed at a certain value and the reflected intensity is recorded as a function of the time. Switching on and off the X-rays, changes in the concentration of color centers can be followed. Figure 5.10 shows the kinetics correspond to the formation and fading of defects in the soda-lime substrates, switching on and off the X-rays at E = 11.95 keV. It is possible to observe that the reflectivity decreases upon irradiation and partially recovers with time after irradiation, varying the concentration of defects.

The measurements of reflectivity as function of time can be carried out without any metallic film, by measuring the light transmission across a glass slide when it is irradiated. However, the main advantage of measuring with this experimental setup (SPR-XAS device) is that it is possible to study simultaneously the kinetics and variations of the refractive index by an analysis of SPR curves.

The setup can also be used to study the effect of X-ray irradiation on organic and inorganic dielectric layers grown on top of the metallic film, since SPR spectroscopy

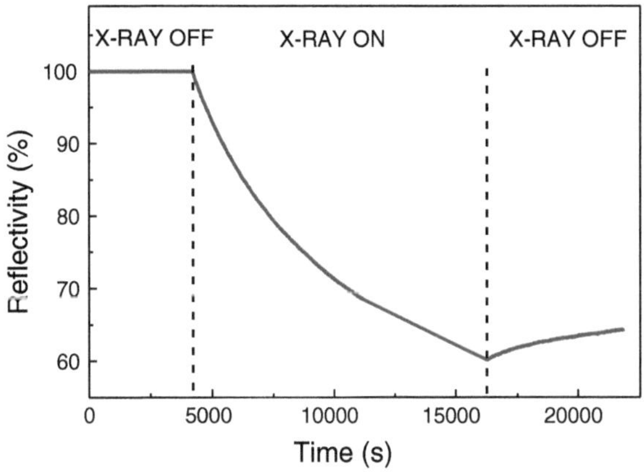

Fig. 5.10 Evolution of reflectivity versus time at an incident angle of 45.6° for a Au thin film grown on soda-lime glass, with and without irradiating with X-rays at E = 11.95 keV

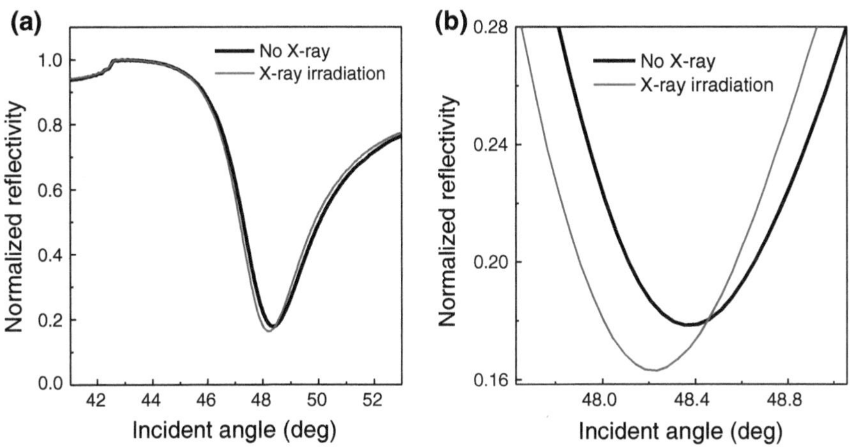

Fig. 5.11 **a** SPR curves for a 5 nm CoPc layer on a 50 nm Au thin films grown on a soda-lime substrate, before and after X-ray irradiation for 360 min using X-rays at E = 7.720 keV (at the Co K edge). **b** A detail of the resonance region

is extremely sensitive to the features of the dielectric medium over the metallic film [2, 10, 17]. Figure 5.11 shows the SPR spectra for a 5 nm Co-phthalocyanine (CoPc) layer grown at 200 °C on a 50 nm Au film deposited onto a soda-lime substrate, measured before and after X-ray irradiation for 360 min using X-rays at E=7.720 keV (at the Co K edge). Upon irradiation with X-rays, a decrease of the intensity around 0.01, a shift towards lower angles at the resonance around 0.2° and a widening of the SPR curve about 0.04° at FWHM are observed (see Fig. 5.11b). These changes are related to the damage induced by the X-rays on the CoPc layer, modifying its refractive index. The same experiment was repeated in a sample with identical glass

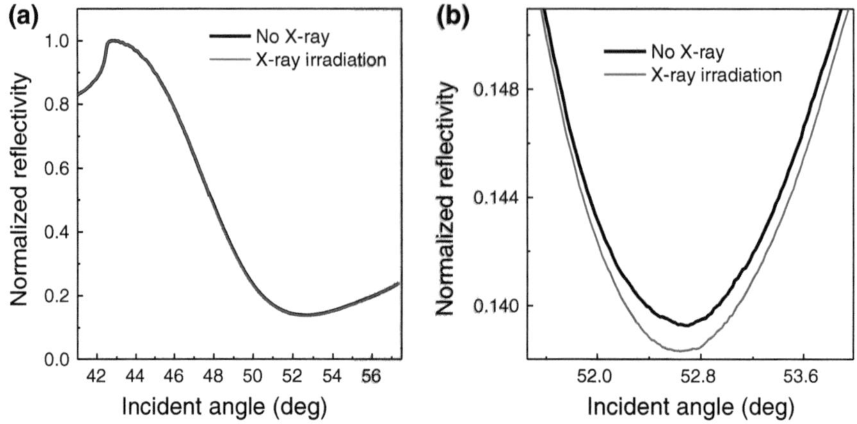

Fig. 5.12 **a** SPR curves for a 5 nm FeO_x layer on a 45 nm Au thin films grown on a silica substrate before and after the X-ray irradiation, for 330 min using X-rays at E = 7.112 keV (at the Fe K edge). **b** A detail of the resonance region

substrate and Au film but without CoPc layer, and this phenomenology did not show up, confirming that the changes of SPR curve are due to modifications of the CoPc film.

Figure 5.12 presents the SPR spectra for a 45 nm Au/5 nm FeO_x[3] bilayer on a silica substrate before and after the X-ray irradiation, for 330 min of irradiation, at E=7.112 keV (at the Fe K edge). Upon X-ray irradiation, no significant variations are observed in the reflected intensity, but a detail of the resonance region (see inset in Fig. 5.12b) shows a small decrease of the reflectivity about 0.001 and a wider resonance around 0.1° at FWHM. A similar sample without the FeO_x layer was analyzed and changes of the same order than with the FeO_x layer were found. Thus, the observed modifications can be related to the Au thin film on silica substrate (see Fig. 5.9b) and not to the FeO_x film.

Therefore, with these experiments we can conclude that the SPR-XAS system is prepared to monitor changes induced by X-ray irradiation in the SPR spectra, tracking both the real and imaginary part of the refractive index of different layers of the samples and studying the kinetics of the process with great accuracy. A detailed study of soda-lime and silica substrates and CoPc films irradiated with X-rays is shown in Chaps. 6 and 7, respectively.

[3] FeO_x layer was prepared by evaporation of a 5 nm Fe film plus annealing in air at 100 °C for 1 h.

5.4.2 *Experimental Setup to Study Instantaneous Effects*

The examples above described correspond to modifications in the structure of the material induced by the X-rays, which are accumulative and mostly permanent. However, this setup allows also studying changes in the electronic structure (modifications in the electron population at the Fermi level) due to X-ray absorption, which are instantaneous and reversible. To achieve the best possible resolution and separate the instantaneous effects from the permanent ones, the configuration illustrated in Fig. 5.13 is designed, which is a modification of experimental setup shown in Figs. 5.1, 5.2 and 5.3. In this case, the laser reaches the sample without any modulation (without chopper) while the X-rays are modulated with a chopper. The signal generated by the reflected laser beam at the photodiode is collected with the lock-in amplifier using as reference that of the chopper. Since the laser is not modulated, in absence of X-rays, the signal measured at the lock-in is zero. However, if the X-rays induce any instantaneous and reversible modification on the material, the reflected laser beam would exhibit a component with the frequency of the chopper that can be detected by the lock-in.

Using this modified setup, possible electronic effects are studied for a Au film grown on soda-lime glass when irradiating with X-rays at the Au L3 edge (11.919 keV). As shown Fig. 5.14, there is no detectable effect up to a relative value of 10^{-5} of the SPR signal, which is the sensitivity of this setup (the background value is independent of the X-ray irradiation). This result is consistent with

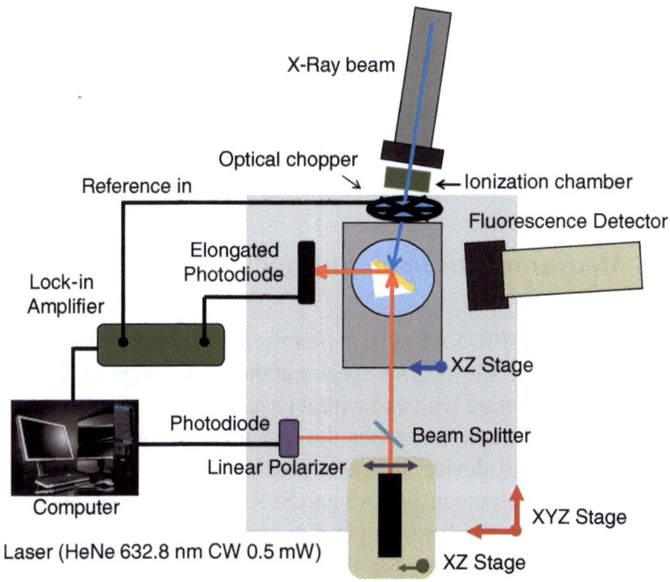

Fig. 5.13 Configuration used to detect instantaneous changes induced by X-rays on the SPR spectrum. Note the different position of the chopper with respect to that of Fig. 5.1

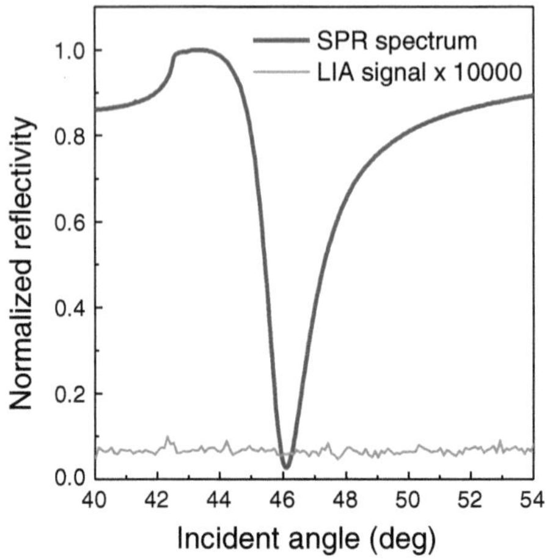

Fig. 5.14 SPR curve for a Au thin film onto soda-lime glass and signal detected by the lock-in amplifier in the configuration shown in Fig. 5.13 (multiplied ×10000)

calculations on the conduction electron density in Au. Considering the X-ray photon flux of the beamline (10^{12} photons/s), the density of conduction electrons for Au of $5.9{\cdot}10^{22}$ electrons/cm^3 [18], the irradiated volume about $5 \cdot 10^{-8}$ cm^3 and the decay time of photoexcited electrons in the conduction band of the order of 10-100 ns, we can estimate that the relative change of conduction electrons in the Au film is of the order of 10^{-12}. Taking into account that the plasma frequency for Au [19] is $\omega_p = 2.18{\cdot}10^{15}$ s^{-1} and the wavelength of used laser is 632.8 nm (frequency $4.74 \cdot 10^{14}$ s^{-1}), we estimate a relative change in the dielectric permittivity of the order of 10^{-12}. This variation can induce changes in the reflected intensity of SPR curve of $\sim 10^{-10}$ (see Appendix A), too small to be detected here. Nevertheless, the achieved resolution (10^{-5}) could be enough to observe changes in other materials with reduced density of electrons at the Fermi level as it is shown below (see Fig. 5.16).

5.4.3 XAS Measurements upon Surface Plasmon Excitation

As indicated above, the SPR-XAS system permits also recording XAS spectra while exciting SPR. Figure 5.15 shows XAS spectra at the Au L3 edge for a 45 nm Au film grown on soda-lime substrate with and without exciting SPR. No difference is found up to 10^{-2}, which is the same order than the resolution of this setup (see Fig. 5.8).

With this experimental device, it is also possible to carry out combined experiments by measuring simultaneously XAS and SPR signal while scanning parameters as time, X-ray energy or incidence angle. For instance, XAS spectra upon excitation of SPR, collecting simultaneously the reflectivity signal as a function of X-ray energy, can be measured. To illustrate this procedure (Fig. 5.16), we have recorded simultaneously the XAS signal and the SPR reflectivity at the resonance condition

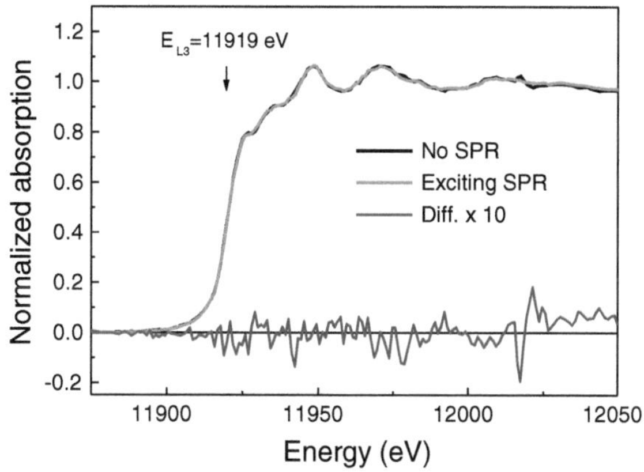

Fig. 5.15 XAS spectra measured at the Au L3 edge on a 45 nm Au film with and without excitation of SPR and their difference

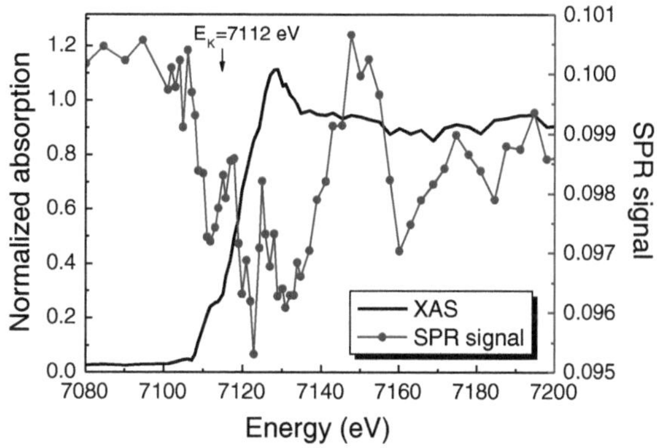

Fig. 5.16 XAS spectra measured at the Fe K edge for a 45 nm Au/ 5 nm FeO_x bilayer grown over a silica glass upon excitation of SPR (*black line*) and the reflectivity signal of the SPR curve at the resonance during the energy scan (*blue circles*)

for the 45 nm Au/5 nm FeO_x bilayer grown on a silica substrate, while scanning the X-ray energy across the Fe K edge (7.112 keV)[4]. A slight decrease can be observed in the SPR signal when the energy crosses the Fe K edge. This feature is related to the X-ray absorption of the FeO_x layer although the mechanism is still not clear: it could be either due to a change in the electron population around the Fermi level or be associated with a local increase of temperature.

5.5 Conclusions

In summary, we have designed and developed a setup combining surface plasmon resonance and X-ray absorption spectroscopies: the SPR-XAS setup. With this experimental device, SPR spectroscopy can be used to explore the interaction of X-rays

[4]The profile of the XANES spectrum suggests that it is not a pure iron oxide film. The sample was prepared by evaporation of a 5 nm Fe film plus annealing in air at 100 °C for 1 h. Thus, apart from a mixture of iron oxides, we cannot discard the presence of some metallic iron in the sample.

with matter and similarly, XAS can be used to study modifications of materials while surface plasmons are excited. Moreover, the system versatility allows measuring simultaneously SPR excitation and X-ray absorption while scanning different parameters. The resolution of the measurements in this device is of the order of 10^{-2} to 10^{-5}, depending on the particular experiment. This setup is currently available for future experiments at the ESRF beamline BM25A-SpLine.

References

1. R.P.H. Kooyman, L.M. Lechuga, Immunosensors based on total internal reflectance. In: *Handbook of Biosensors and Electronic Noses: Medicine, Food and the Environment* (CRC Press, Boca Raton, 1997)
2. W. Knoll, Interfaces and thin films as seen by bound electromagnetic waves. Annu. Rev. Phys. Chem. **49**, 569–638 (1998)
3. D.C. Koningsberger, *R. Prins, X-ray Absorption: Principles, Applications, Techniques of EXAFS, SEXAFS and XANES* (Wiley, New York, 1987)
4. D. Norman, X-ray absorption spectroscopy (EXAFS and XANES) at surfaces. J. Phys. C: Solid State Phys. **19**(18), 3273 (1986)
5. R. Livingston, C.E. Nurnberger, A kinetic interpretation of the coloring of glass by x-rays. J. Phys. Chem. **39**(7), 1011–1020 (1934)
6. E. Itälä, Fragmentation processes in organic molecules induced by synchrotron radiation. PhD thesis, University of Turku, Finland, 2012
7. A. Serrano, O. Rodríguez de la Fuente, V. Collado, J. Rubio-Zuazo, C. Monton, G. Castro, M. García, Simultaneous surface plasmon resonance and x-ray absorption spectroscopy. Rev. Sci. Instrum. **83**(8), 083101 (2012)
8. See http://www.esrf.eu/UsersAndScience/Experiments/CRG/BM25
9. G.R. Castro, Optical design of the general-purpose Spanish x-ray beamline for absorption and diffraction. J. Synchrotron Radiat. **5**(3), 657–660 (1998)
10. H. Raether, *Surface Plasmons on Smooth and Rouge Surfaces and on Gratings* (Springer, Berlin, 1988)
11. H. Raether, Surface plasma oscillations and their applications, vol. 9. In: *Physics of Thin Films*, ed. by G. Hass (Academic, New York 1977)
12. H. Raether, E. Kretschmann, Radiative decay of non radiative surface plasmons excited by light. Zeitschrift für Naturforschung A **23**, 2135–2136 (1968)
13. A. Otto, Excitation of nonradiative surface plasma waves in silver by the method of frustrated total reflection. Zeitschrift für Physik **216**, 398–410 (1968)
14. C.E. Nurnberger, R. Livingston, Further studies on the kinetics of the coloring of glass by x-rays. J. Phys. Chem. **41**(5), 691–697 (1937)
15. D.L. Griscom, Optical properties and structure of defects in silica glass. J. Ceram. Soc. Jpn. **99**(1154), 923–942 (1991)
16. J. Sheng, Y. Wu, X. Yang, J. Zhang, UV-laser irradiation on the soda-lime silicate glass. Int. J. Hydrogen Energy **34**(2), 1123–1125 (2009)
17. J. Homola, S.S. Yee, G. Gauglitz, Surface plasmon resonance sensors: review. Sens. Actuators A Chem. **54**, 3–15 (1999)
18. C. Kittle, *Introduction to Solid State Physics*, 8th edn. (Wiley, New York, 2005)
19. M.A. Ordal, L.L. Long, R.J. Bell, R.S. Bell, R.R. Bell, J.R.W. Alexander, C.A. Ward, Optical properties of the metals Al Co, Cu, Au, Fe, Pb, Ni, Pd, Pt, Ag, Ti, and W in the infrared and far infrared. Appl. Opt. **22**, 1099–1119 (1983)

Chapter 6
In-Situ Study of Glasses Irradiated with X-rays by Surface Plasmon Resonance

This chapter presents a study of the effects of hard X-ray irradiation on glasses carried out using the SPR-XAS device described in the previous chapter. Two types of glasses are analyzed: soda-lime and silica substrates. Information on variations of the refractive index of glasses upon irradiation and the kinetics of the induced defects are obtained.

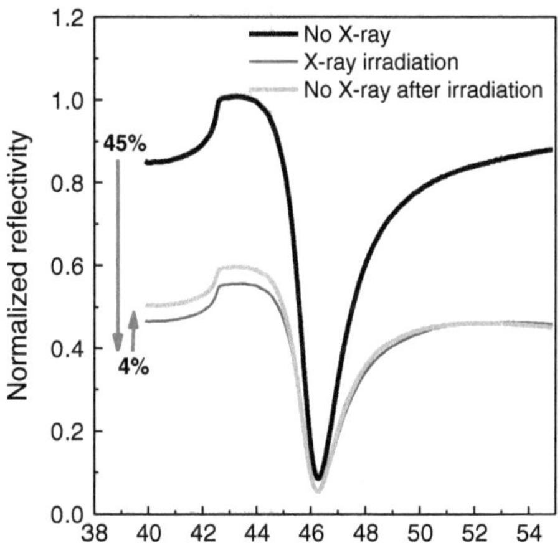

6.1 Introduction

Glassy materials present very appealing optical and electrical properties for various industrial applications including optical fibers, optical components, hot-cell windows or reactors. Their advantages mainly rely on low cost and high optical transparency, which extends from the infrared to ultraviolet regions [1–4]. High energy ionizing radiation (X-rays, gamma rays, electrons, etc.) can induce numerous changes in

A. Serrano Rubio, *Modified Au-Based Nanomaterials Studied by Surface Plasmon Resonance Spectroscopy*, Springer Theses, DOI 10.1007/978-3-319-19402-8_6

glasses, being the visible coloration one of the most obvious [1, 5–7]. In particular, X-ray irradiation can induce color centers in some glasses [8, 9], leading to the formation of these defects responsible of the glass darkening [10] and prompting a great interest since these color centers can be generated and bleached reversibly [11]. These and other non-visible defects generated with radiation have been investigated in a wide variety of glassy materials in the last decades, being responsible of the alteration of some physical properties of the glasses such as the optical absorption [8, 11], refractive index [12], axial stress [13] or thermal expansion coefficient [5].

It is often necessary that glassy materials properly respond under radiation [1, 4] for a specific application. Therefore, studies on defects in glasses induced by X-ray irradiation are important to determine their nature, the possible mechanisms during and after the radiation and the changes on the glass properties. In the literature, some works have studied the formation and fading dynamic of color centers and/or their possible effects on physical properties when these defects are induced [1, 5, 6]. Most of these studies devoted to analyze the effect of X-ray irradiation onto glasses, are carried out by irradiating the sample and analyzing later their optical properties [8–11]. However, in situ and real time studies are necessary for a deeper understanding of the process. Otherwise, processes associated with fast dynamics can be missed. Hence, measuring the optical properties of the glasses in situ and in real time while the glass is irradiated and immediately after irradiation, results appealing. The latter study can contribute to the understanding of fast processes related to the healing of the defects produced. Performing ellipsometric measurements in combination with X-ray irradiation is complicated. Optical absorption spectroscopy can be easier to combine with X-ray irradiation, but it just rends information about the imaginary part of the refractive index. However, surface plasmon resonance (SPR) spectroscopy provides simultaneously information on the modification of the real and imaginary parts of the refractive index. As previously explained, SPR in metallic films is very sensitive to the features of the surrounding media, and thus to the substrate [14]. Measuring SPR curves of metallic films grown on glass substrates while irradiating with X-rays allows tracking variations in the real and imaginary parts of the refractive index of the glass in real time [15].

In this chapter, we present a study of SPR on Au thin films/glass substrates upon X-ray irradiation. Using the SPR-XAS setup [16], described in Chap. 5, we can study in situ and real time the effects of X-ray irradiation on the glasses by SPR [15]. Two types of glass substrates are analyzed: soda-lime and silica substrates. On the one hand, soda-lime glass presents a structure with a high concentration of non-bridging oxygen defects [17] and they show a darkening when are irradiated with X-rays. On the other hand, silica glass shows a negligible concentration of non-bridging oxygen defects [17] and they do not present visible defects when are irradiated with X-rays. Preliminary results of the effect of X-rays on glasses were presented in Chap. 5.

6.2 Fabrication of Samples

Au thin films with a thickness around 40 nm were grown onto soda-lime and silica glass substrates by thermal evaporation. A detailed description of the glass composition is shown in Sect. 2.2.2 of Chap. 2. Soda-lime and silica substrates were cleaned prior to deposition with soap and water, and dried with dry air flux. The thermal evaporation of the Au thin films was performed on the glass substrates using the home-made evaporation chamber mounted in the Departamento de Electrocerámica at the Instituto de Cerámica y Vidrio (CSIC) of Madrid (see Chap. 2, Fig. 2.4). For the evaporation of Au metallic films, substrates were placed about 15 cm away from the filament and the deposition was performed under a 10^{-6} torr pressure with currents of 26 A at a rate of 0.02 nm s^{-1}, controlled with a Q-microbalance.

6.3 Experimental Details

Experiments were performed at the branch A of the BM25 SpLine beamline at the European Synchrotron Radiation Facility (ESRF) in Grenoble (France), with the SPR-XAS device [16], previously described in Chap. 5. Briefly, this experimental setup follows the Kretschmann-Rather configuration [18] and allows the simultaneous measurements of SPR and XAS and studying the SPR in situ while the sample is being irradiated. For measuring SPR upon X-ray irradiation, samples were simultaneously illuminated with the laser from the prism side and the X-rays from the Au film side, making the spots coincident at the Au film side (see Chap. 5, Figs. 5.1, 5.2 and 5.4). Although the X-ray beam reaches the Au film, the skin depth for hard X-rays in Au is of the order of several microns, so X-rays pass through the Au thin film reaching the substrate at the same position as the laser spot.

Simulations of SPR spectra were performed with the freeware software *Winspall* [19]. Numerical fits were carried out manually in order to control the modified layers by X-rays. Optical absorption spectra were recorded using the V-670 UV-Visible double beam spectrophotometer.

6.4 Study of Soda-Lime Glasses by SPR-XAS Setup

Au thin films grown onto soda-lime glass substrates were irradiated with X-rays around the Au L3 edge at different energies: 11.80, 11.92 and 11.95 keV, which correspond to energies below, at and above the absorption edge, respectively.

Figure 6.1a–c shows experimental SPR spectra for a 42 nm Au film grown onto a soda-lime substrate glass before, during and after X-ray irradiation at 11.80, 11.92 and 11.95 keV, respectively. In all cases, the irradiation induces a decrease in the intensity of the reflected beam for the whole SPR curve. This behavior is due to the darkening

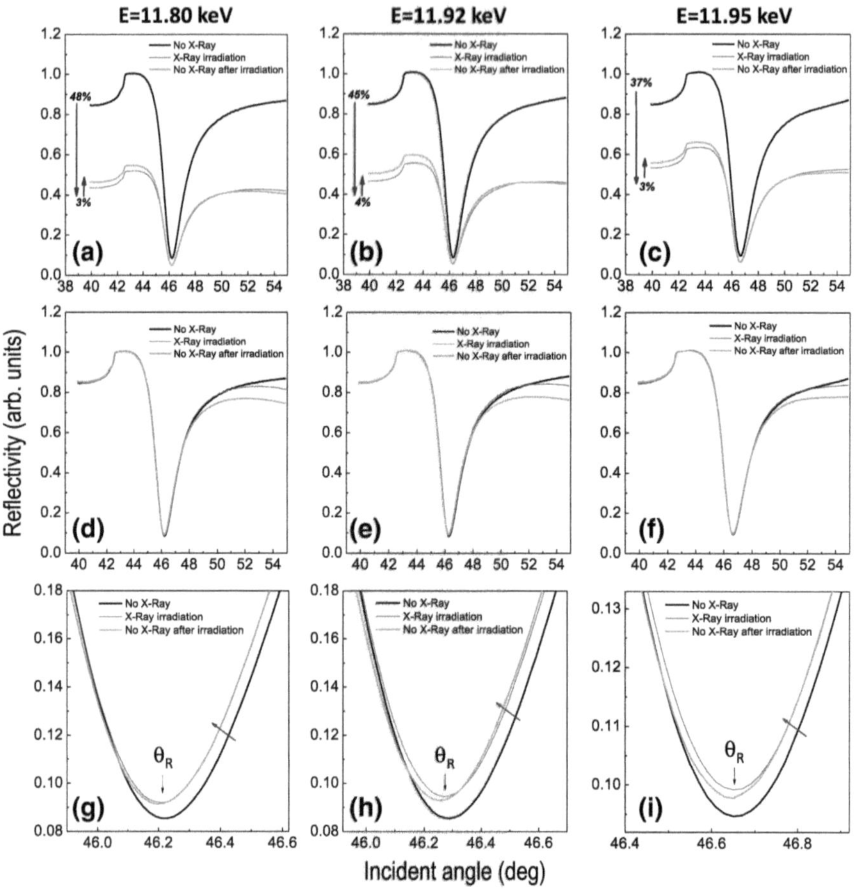

Fig. 6.1 **a–i** SPR curves for a 45 nm Au film on soda-lime glass before (*black line*), during (after 230 min of X-rays, *blue line*) and after (after 170 min without X-rays, *red line*) an irradiation with X-rays at 11.80, 11.92 and 11.95 keV, respectively. **a–c** Spectra as measured, **d–f** spectra normalized at the critical angle (42.61°) and **g–i** a detail of the resonance region for the normalized spectra

of the glass upon X-ray irradiation, which is observable with the naked eye. When measuring SPR, the laser beam propagates through the glass substrate and is partially absorbed reducing its intensity. The effect is accumulative: the larger the irradiation time, the larger the decrease of the intensity. SPR spectra intensity decreases about 48, 45 and 37 % at 11.80, 11.92 and 11.95 keV respectively (see Table 6.1), after 230 min of X-ray irradiation. When switching off the X-rays, the reflected intensity of SPR spectrum increases, pointing out that the irradiation effects are partially reversible: after 170 min without X-ray irradiation, the SPR spectra intensity increases from 48 to 51 % at 11.80 keV, from 45 to 49 % at 11.92 keV and from 37 to 40 % at 11.95 keV (see Table 6.1). The glass darkening is due to the formation of color centers upon irradiation. These defects are unstable at room temperature, so the darkening is

Table 6.1 Intensity variation between SPR spectra measured before, during and after X-ray irradiation at 11.80, 11.92 and 11.95 keV of Au films grown on soda-lime glasses

Reflectivity variation					
E (keV)	No X-ray irradiation	⇒	X-ray irradiation	⇒	After X-ray irradiation
11.80		230 min	 ↓ 48 %	170 min	 ↑ 3 %
11.92		230 min	 ↓ 45 %	170 min	 ↑ 4 %
11.95		230 min	 ↓ 37 %	170 min	 ↑ 3 %

partially reduced with time after irradiation. However, the initial reflectivity values are not completely recovered with time (months are necessary) [8]. Figure 6.1d–f shows the normalized reflected intensity of the SPR spectra measured before, during and after irradiation, respectively. We can observe that the shape of SPR curve changes at larger incident angles when the sample is irradiated. This effect is not recuperated when the X-rays are switched off but it is even deeper.

The resonance angle of SPR spectra is extremely sensitive to any change on the refractive index of the glass substrate. Upon irradiation, the resonance angle slightly shifts towards lower angles and the full width half maximum (FWHM) is somehow reduced. These effects increase with irradiation time. For 230 min irradiation, the shift is about 0.035° at 11.80 keV (Fig. 6.1g), 0.040° at 11.92 keV (Fig. 6.1h) and 0.025° at 11.95 keV (Fig. 6.1i). Neither the shift nor the narrowing of the resonance is reversible after irradiation.

The critical angle does not change upon X-ray irradiation since it depends just on the refractive index of the quartz prism and the air (see Fig. 6.3d). As indicated in Chap. 2, total reflection takes place at the substrate/Au interface. When the laser passes from the prism to the soda-lime glass, according to Snell's law, the refraction keeps the component of the light wavevector parallel to the surface constant in both media. Since the critical angle is determined by this component, it will not depend on the particular value of the refractive index of the soda-lime glass.

It is well known that the refractive index in glasses increases by exposing them to X-ray radiation [12] or UV laser pulses [20]. The glass presents, in this case, a profile density of the induced defects which is maximum at the Au/glass interface and decreases exponentially inside the glass (see Fig. 6.3d). We could observe that the darkening comes across the whole glass, being even transmitted (but with low intensity) into the prism. However, simulations of the SPR spectra result too complicated if this gradient is fully taken into account. Instead, a homogeneous darkening along the whole soda-lime substrate was considered, which has a thickness of 1 mm (see Fig. 6.3h). This approach has allowed studying semi-quantitatively the modifications of the glass induced by the X-rays in the optical properties of the glass. Therefore,

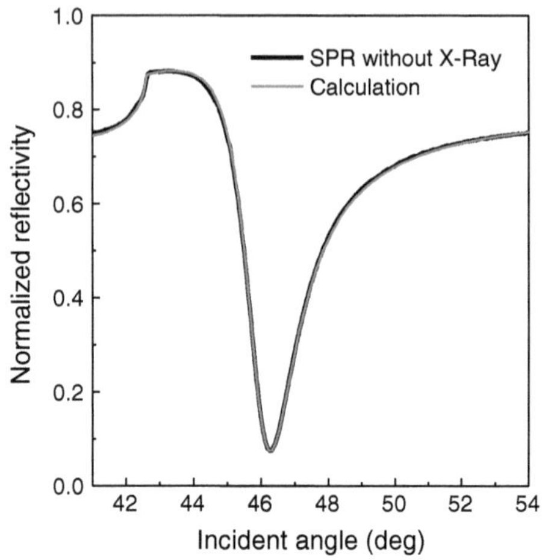

Fig. 6.2 Experimental (*black line*) and calculated (*red line*) SPR curves of a 42 nm Au film on soda-lime glass substrate without X-ray irradiation. The used fit parameters are: $n_{prism} = 1.457$ and $n_{Au} = 0.1425 + 3.53\ i$ at $\lambda = 633$ nm

the obtained values for the refractive index are approximated and show a tendency rather than an exact value.

In order to achieve the proper performance of the SPR simulations, we initially fitted the SPR spectrum for a 42 nm Au film grown onto soda-lime glass with the X-ray switched off, as shown Fig. 6.2. The fit of SPR spectrum was achieved using $n_{prism} = 1.457$ and $n_{Au} = 0.1425 + 3.53\ i$ at $\lambda = 633$ nm. With these parameters, we can observe a good agreement between the measured and simulated SPR spectra.

Subsequently, simulations of measured SPR were performed varying the refractive index of the modified layer, in a four-media system, in order to reproduce the experimental results presented in Fig. 6.1. The simulations were carried out for measured SPR curves of a 42 nm Au film on soda-lime glass irradiated with X-rays at 11.92 keV (Fig. 6.1b, e, h). In these simulations, two glass media were considered: an infinite medium described with the parameters of the unmodified quartz prism (i.e., $n_{prism} = 1.457$) plus a 1 mm layer (soda-lime glass substrate) with modified properties that are indicated in Fig. 6.3. The refractive index of soda-lime glass upon X-ray irradiation was determined by fitting the experimental SPR curve. In this fit, the refractive index of the soda-lime glass is the only free parameter, while the rest of parameters remain fixed with the same values as those prior to X-ray irradiation.

Figure 6.3 shows the results of measured SPR curves [see Fig. 6.3a (6.1b), 6.3b (6.1e) and 6.3c (6.1h)] and the corresponding fits (see Figs. 6.3e–g). A very good agreement with the experimental results can be obtained assuming that the refractive index of the modified layer upon irradiation changes from $n_{SL} = 1.513$ (non-irradiated value) to $n_{SL} = 1.613 + 1.5010^{-5}i$, (see Fig. 6.3e–g). To reproduce the fine changes in the structure of the resonance it is necessary to consider also a slight modification of the Au effective refractive index of the order of 1 %. A modification of the Au intrinsic refractive index induced by the X-ray irradiation is not likely. The modification might be rather due to structural or electric modifications at the glass/Au interface upon irradiation. This could be also due to the adsorption, induced

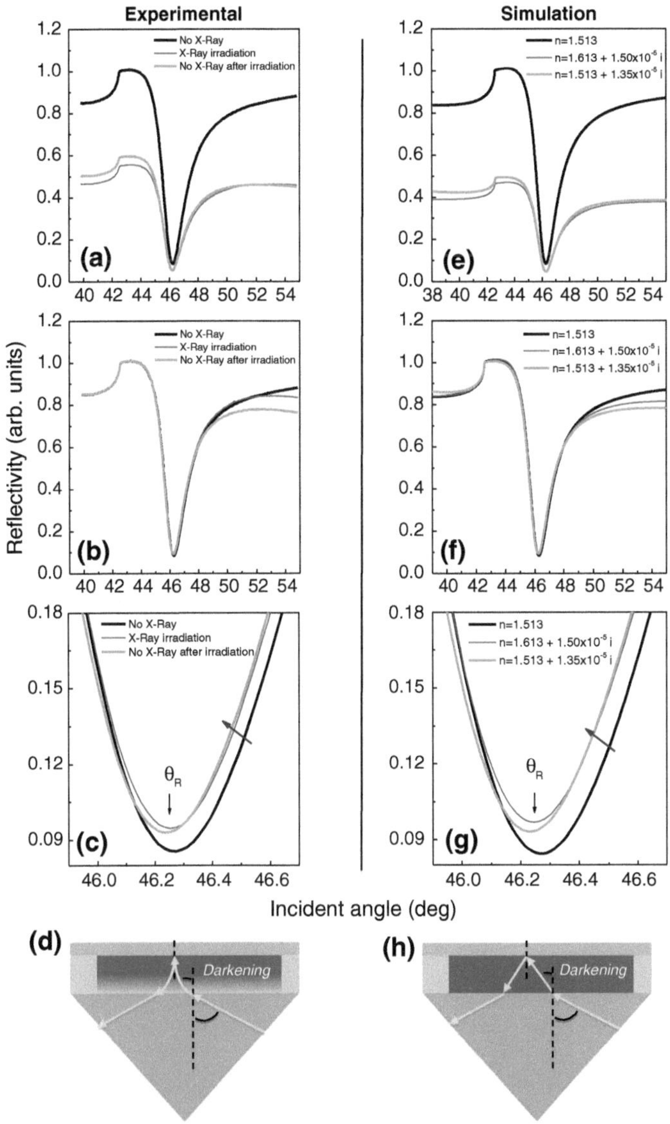

Fig. 6.3 **a–c** SPR curves for a Au film on soda-lime glass before (*black line*), during (*blue line*) and after (*red line*) irradiation with 11.92 keV X-rays. **a** Spectra as measured, **b** spectra normalized at the critical angle and **c** a detail of the resonance region for the normalized spectra. **e–g** Simulated SPR spectrum using the parameters indicated in the legend. **e** Unnormalized spectra, **f** spectra normalized at the critical angle and **g** a detail of the resonance region for the normalized spectra. **d** and **h** Represent a scheme of the four-media system (*darkening* is decreasing inside the soda-lime glass) and the model used for the simulations (*darkening* is homogeneous), respectively

by irradiation, of molecular or atomic species or impurities at the surface of Au, such as atomic oxygen, when the surface is eventually exposed to the ozone [21] produced by synchrotron radiation in the experimental hutch. Note that, with the reported values, it is possible to reproduce simultaneously the reduction in the intensity, the change of the spectrum profile at large incident angles, the shift of the resonance angle towards smaller values and the decrease of the FWHM (compare experimental results with simulations in Fig. 6.3). Regarding the partial recovery of the SPR signal after irradiation, the observed behavior may be reproduced assuming that the modification of the real refractive index is totally recovered, while the modification of imaginary one is almost permanent (recovery of only a 10 % after 170 min, for sample irradiated with X-rays at 11.92 keV).

As mentioned above, soda-lime glass darkening upon X-ray irradiation is due to the generation of color centers [8]. These color centers are attributed to the formation of electron-hole pairs which, in turn, can yield the creation of defects, causing high absorbance in the UV and visible region. According to the literature [1, 22, 23], radiation induces defects on glasses associated with either oxygen excess or oxygen defect in the network. For soda-lime glass, the most typically reported defects are non-bridging oxygen hole center (NBOHC: ≡Si–O*), E' center (≡Si*), peroxy radical (POR: ≡Si–O–O*) and trapped electron (TE), where the symbol "≡" denotes three bonds with other oxygens in the glass network and the "*" represents an unpaired electron [1, 22].

A proper identification of the type of color center can be obtained by optical absorption spectroscopy, since each of these defects induces absorption bands at different wavelengths [8, 9]. Figure 6.4 shows the result of an irradiated soda-lime

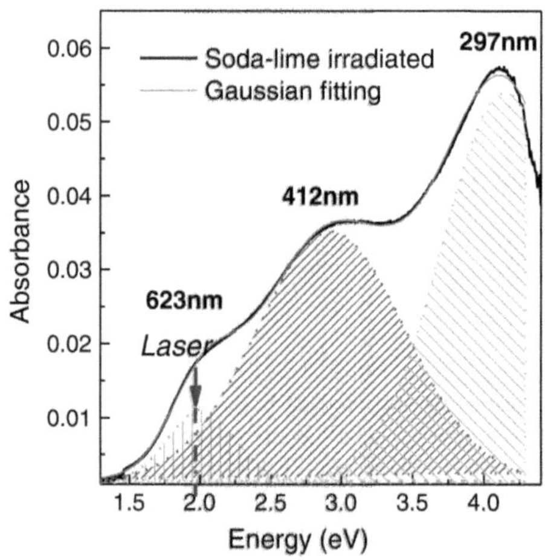

Fig. 6.4 Optical absorption spectrum for a soda-lime glass irradiated at E = 11.80 keV (*black line*) and the fit (*red line*) to three gaussian bands associated with the induced absorptions. The measurement was performed one month after X-ray irradiation. The bands at 623 and 412 nm correspond to NBOHC defects and the band at 297 nm corresponds to TE defects. The wavelength of used laser in the SPR experiments (632.8 nm) is indicated in the figure

Table 6.2 Parameters corresponding to the three gaussian fits of the optical absorption bands for soda-lime glass irradiated with X-rays at E = 11.80 keV

Wavelength (nm)	Energy (eV)	FWHM (eV)	Defect
623.3 ± 0.2	1.990 ± 0.001	0.385 ± 0.002	*HC1*
411.7 ± 0.3	3.012 ± 0.002	1.347 ± 0.002	*HC2*
297.3 ± 0.1	4.171 ± 0.001	0.765 ± 0.002	*TE*

glass (referenced to non-irradiated soda-lime glass in order to obtain the net irradiation induced absorption) measured 1 month after the irradiation. The spectrum can be decomposed into three gaussian bands (with a coefficient of determination $R^2 = 0.99977$). In this way, the three characteristic absorption bands with maxima around 623, 412 and 297 nm (2.0, 3.0 and 4.2 eV, respectively) are resolved. Table 6.2 summarizes the main features of these absorption bands.

According to the literature, the bands at 623 and 412 nm are associated with the induced defects of NBOHCs [1, 22], being the responsible of the brown color, while the band of 297 nm is identified as absorption of TEs [8]. There are two types of NBOHC defects, designed as HC1 and HC2, which are paramagnetic and dominate the optical absorption spectra of glasses with low and high alkali contents, respectively [24]. The band at 623 nm (HC1) corresponds to a hole trapped in the 2p orbital of one non-bridging oxygen at those sites of the network where the SiO_4 polyhedron contains three bridging oxygen and one non-bridging oxygen, all bonded to a silicon atom. HC2 defects are responsible of the absorption band around 412 nm, which is associated with a hole trapped in two or three non-bridging oxygens bonded to the same silicon [11, 25, 26]. In the SPR experiments a 632.8 nm laser was used, and thus the observed absorption at this wavelength is due to both HC1 and HC2 defects (see Fig. 6.4).

The generation of NBOHC defects in a soda-lime glass upon X-ray irradiation starts with the formation of electron-hole pairs. In silica-based glasses, the holes are trapped by the non-bridging oxygen atoms yielding the formation of NBOHCs, while electrons are diffused through the glass network until finding holes for recombining. Since these NBOHC defects are unstable at room temperature, they can be eliminated after irradiation by recombination with electrons or other NBOHCs according to the following reactions [1, 8]:

$$\equiv\mathrm{Si{-}O}^* + e^- \rightarrow \mathrm{Si{-}O}^- \tag{6.1}$$

$$2\,(\equiv\mathrm{Si{-}O}^*) \rightarrow \mathrm{Si{-}O{-}O{-}Si}\equiv \tag{6.2}$$

$$2\,(\equiv\mathrm{Si{-}O}^*) \rightarrow \mathrm{Si{-}O{-}O{-}Si}\equiv + \frac{1}{2}\mathrm{O}_2 \tag{6.3}$$

Paramagnetic defects as NBOHCs decay obeying first order (i.e., an exponential function) or second order kinetics (i.e., the inverse of defect concentration versus the time is linear) [8, 27]. In particular, when the decay is due to recombination with electrons (Reaction 6.1), a first order kinetics is followed. On the contrary, NBOHCs that recombine with each other (Reactions 6.2 and 6.3), exhibit a second order kinetics. Previous studies on X-ray irradiated soda-lime glasses demonstrated that the fading of color centers after irradiation presents a first order kinetics [8, 11, 28, 29]. However, these studies have been performed ex situ and measured over long times (days and/or months). Hence, the study of the kinetics in the order of seconds and/or minutes, which we can monitor with our setup, is still missing.

Using the SPR-XAS device (described in Chap. 5), the reflectivity was measured for a fixed incident angle close to the resonance (45.6°) when switching on and off the X-rays. These measurements could, in principle, have been carried out without the Au thin film or in transmission mode using the same setup. Figure 6.5 shows the effect of X-ray irradiation on soda-lime glass at three different energies, 11.80, 11.92 and 11.95 keV, which are below, at and above the L3 edge of Au, to discriminate if part of the effects are due to X-ray absorption by the Au film. The results confirm that the reflectivity decreases upon irradiation and partially recovers with time after irradiation. Moreover, it is possible to observe that the kinetics of the process depends on the X-ray energy. In particular, the decay results slower when using X-rays well above the Au L3 edge. These differences in decay rate cannot be associated with the reduction of X-ray intensity when passing through the Au film. Only 1 % of the photons are absorbed in the film. However, the X-ray absorption in the film creates photoelectrons that could be ejected towards the glass modifying the lifetime of the defects.

The fitting of these curves to a first order kinetics or to a second order kinetics was not satisfactory. However, as Fig. 6.6 illustrates, these data can be properly fitted to the sum of two exponentials, according to the following equation:

$$I = I_0 + a \times e^{-\frac{t}{\gamma_1}} + b \times e^{-\frac{t}{\gamma_2}} \tag{6.4}$$

where γ_1 and γ_2 represent decay constants and I_0, a and b are fitting parameters.

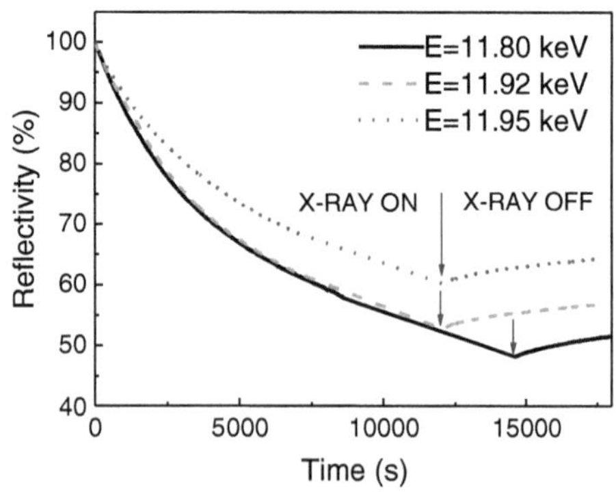

Fig. 6.5 Time dependence of the reflectivity when switching on and off the X-rays at three different energies (11.80, 11.92, and 11.95 keV), in situ and in real time. The X-ray beam at 11.80 keV is switched off 2500 s later than the two other cases

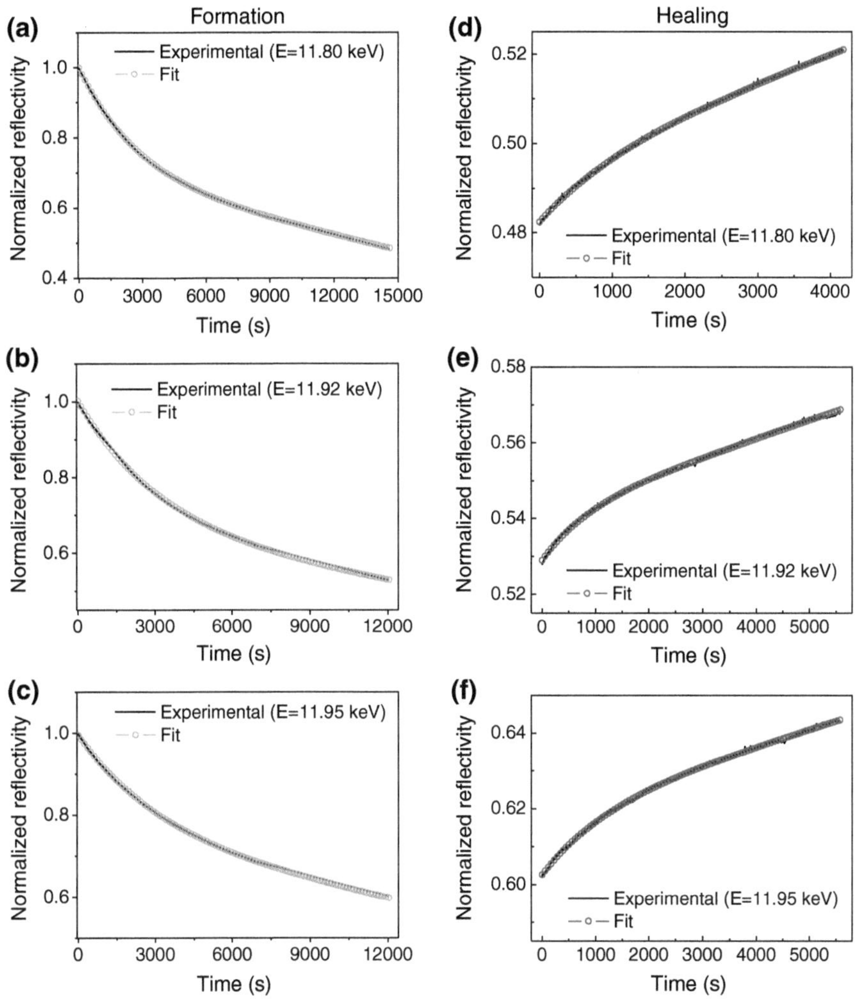

Fig. 6.6 Reflected intensity versus time (*black solid lines*) and fitting of the color centers **a–c** formation (*red circles*) and **d–f** fading (*blue circles*) by switching on and off X-rays at 11.80, 11.92 and 11.95 keV, at an incident angle of 45.6°. Spectra are normalized to the maximum of the reflectivity (100 %)

As a result of the exponential fitting (Eq. 6.4), we conclude that the formation and fading of defects correspond to two first order kinetics, where for each first order reaction the half-life time $\tau_i (i = 1, 2)$ can be calculated from:

$$\tau_i = ln2 \times \gamma_i \tag{6.5}$$

Table 6.3 Parameters corresponding to the exponential fitting for the defects formation (*left*) and fading (*right*) kinetics switching on and off X-rays, respectively, at three different energies (11.80, 11.92, and 11.95 keV)

Defects formation				Defects healing			
E (keV)	11.80	11.92	11.95	E (keV)	11.80	11.92	11.95
I_0	−2720	−1922	−3036	I_0	43	143	180
a	0.31	0.35	0.25	a	−0.01	−0.01	−0.02
$\gamma_1(s)$	2729	3245	3195	$\gamma_1(s)$	1053	877	1192
$\tau_1(s)$	1891	2249	2215	$\tau_1(s)$	730	608	826
b	2721	1922	3037	b	−43	−142	−180
$\gamma_2(s)$	1.93×10^8	1.72×10^8	2.34×10^8	$\gamma_2(s)$	6.94×10^6	2.95×10^7	4.12×10^7
$\tau_2(s)$	1.34×10^8	1.19×10^8	1.62×10^8	$\tau_2(s)$	4.81×10^6	2.04×10^7	2.86×10^7
R^2	0.9999	0.9996	0.9999	R^2	0.9994	0.9987	0.9991

These fittings were carried out for the three energies studied obtaining coefficients of determination better than 0.9987. Table 6.3 shows the parameters corresponding to the fittings for each energy. The decay of the color centers follows double first order kinetics with decay times of the order of minutes and months. The latter is qualitatively in agreement with previous ex situ studies of the irradiation effects [8, 30]. On the contrary, the decay constant of the order of minutes has not been previously reported. The fact that just first order kinetics are found in the fading curves, points out that the NBOHC defects just recombine with the neighboring electrons (Reaction 6.1) but not with another neighboring NBOHC or another holes (Reactions 6.2 and 6.3). The simulations of SPR spectra, presented in Fig. 6.3, show that the modifications in the real part of the refractive index are partially recovered in minutes after irradiation. On the other hand, the changes in the imaginary part of the refractive index remain almost unvaried for hours after irradiation. Thus, these changes in imaginary refractive index could be mainly associated with the slow kinetics.

Regarding the nature of the two time components, a possible explanation could be that each of them is associated with one of the NBOHCs. However, in this case the band associated with the fast kinetics should disappear from the spectrum after a few hours. Instead, the optical bands observe clearly even after several days (see Fig. 6.4). Moreover, Sheng et al. [8, 30] found that the relative intensity of the bands associated with the NBOHCs does not change with time. Therefore, it is possible to conclude that the fading kinetics are governed by the availability of electrons to recombine with the NBOHCs (first order kinetics). The existence of the two components rather supports that there are two types of electrons responsible of the fast and slow kinetics. The electrons responsible of the fast component should be weakly trapped or nearly free so that they can move through the glass at room temperature, reaching the NBOHCs and recombining in minutes. On the other hand, the electrons associated with the slow kinetics should be those trapped in TE defects, well described in the literature [1, 8, 22] and detected by optical absorption (see Fig. 6.4). The availability of these electrons for recombination is not determined by their mobility but by the probability to be released from a TE defect, which at room temperature yields a decay time of months. As the fading kinetics for both NBOHCs is determined by the presence of electrons available for recombination, the decay rate is the same for both NBOHCs and their relative concentration does not change in agreement with previous results [8, 30]. Comparing these results with those of Sheng et al. [8, 30], quantitative differences can be observed. Here a slow first order kinetics is found with a decay time of the order of months, while Sheng et al. [8] reported a slow first order kinetics of the order of a year (strongly dependent on the irradiation dose). A possible explanation for the differences with these experiments could be that Sheng et al. used 50 keV X-rays, while here the experiments are carried out with beams of about 12 keV. However, the optical absorption spectra profiles are very similar in both cases, pointing out that the types of generated defects are the same, despite the different X-ray energy. Our in situ fading measurements are extended for a maximum of 6 h. With this recording time it is difficult to measure accurately decay times of the order of months. Consequently, the slow kinetics data from Sheng et al. seems more reliable that ours. The main innovation of our experiments is the observation of a fast fading component of the order of minutes (see Table 6.3), not previously reported. This component can be only detected by measuring the optical properties in real time at the irradiation site [15].

6.5 Study of Silica Glasses by SPR-XAS Setup

Effects of X-ray irradiation on dielectric layers grown on a metallic film can be studied, as explained in Chap. 5. As indicated above, silica substrates do not present visible defects when they are irradiated with X-rays, thus these may be good candidates for analyzing possible effects of the X-rays on dielectric overlayers grown on metallic films, discarding those of the substrate. In order to check it, we have carried out a similar study to that of the soda-lime substrates. Therefore, Au thin films grown onto silica glass substrates were irradiated with X-rays around the Au L3 edge at different energies: 11.80, 11.92 and 11.95 keV.

Figure 6.7 shows experimental SPR spectra for a 37 nm Au film grown onto a silica substrate glass before, during and after X-ray irradiation at 11.95 keV. In contrast to soda-lime substrates, for this case, no significant variations are found in the reflected intensity upon X-ray irradiation for 170 min. However, a fine detail of the resonance region, which is extremely sensitive to any change on the refractive index of the glass substrate, reveals a widening of the curve around 0.02° at FWHM. This effect is accumulative: the larger the irradiation time, the larger the effect in the SPR spectrum. When the X-rays were switched off, after 40 min without X-ray irradiation, the SPR spectrum was partially recovered (see Fig. 6.7b). Equivalent results were obtained at 11.80 and 11.92 keV.

In order to achieve the proper performance of the SPR simulations, in the same way than for soda-lime glass substrates, initially the SPR spectrum for a 37 nm Au

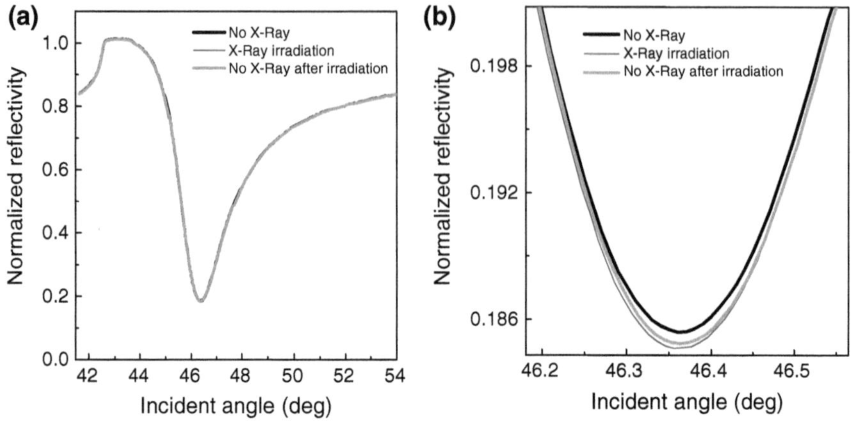

Fig. 6.7 **a** SPR curves for a 37 nm Au film on silica glass before (*black line*), during (after 170 min of X-rays, *blue line*) and after (after 40 min without X-rays, *red line*) irradiation with X-rays at 11.95 keV. **b** A detail of the resonance region

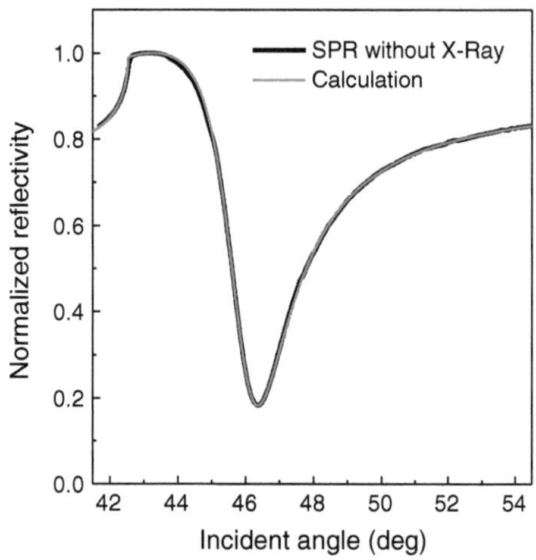

Fig. 6.8 Experimental (*black line*) and calculated (*red line*) SPR curves of a 37 nm Au film on silica glass substrate without X-ray irradiation. The used fit parameters are: $n_{prism} = 1.457$ and $n_{\mathrm{Au}} = 0.14030 + 3.606\,i$ at $\lambda = 633$ nm

film grown onto silica glass was fitted with the X-ray switched off. The fit of SPR spectrum was achieved using $n_{prism} = 1.457$ and $n_{\mathrm{Au}} = 0.14030 + 3.606\,i$ at $\lambda = 633$ nm, obtaining a good agreement between the measured and simulated SPR spectra, shown in Fig. 6.8.

Thereafter, simulations of the measured SPR were performed varying the refractive index of the modified layer, in order to reproduce the experimental results presented in Fig. 6.7. In this case, it was not possible to fit the measured SPR after irradiation modifying the refractive index of the silica glass, in our four-media system. To reproduce the variations in the resonance region, it was necessary to consider some variation where the Au thin film is involved (similar to simulations of soda-lime

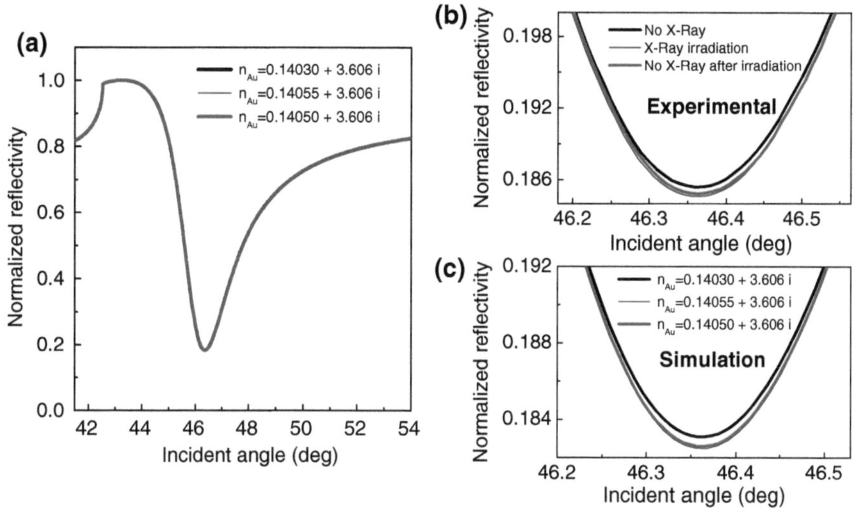

Fig. 6.9 **a** Simulated SPR spectrum using the parameters indicated in the legend for a 37 nm Au film onto silica glass substrates before (*black line*), during (*blue line*) and after (*red line*) irradiation with X-rays at 11.95 keV. **b** A detail of the resonance region for the normalized experimental spectra and **c** a detail of the resonance region for the normalized calculated spectra

glasses). Therefore, for these simulations, we consider the Au thin film as the only modified medium by X-rays, so its refractive index is the only free parameter, while the rest of parameters (included those of the silica glass) remain fixed with the same value as those prior to X-ray irradiation.

Figure 6.9 shows the results of calculated SPR curves and a detail of resonance region compared with the resonance region of measured SPR spectra. In this fit, we can obtain a very good agreement with the experimental results assuming that the real refractive index of the modified layer (n_{Au}) upon irradiation increases about 0.00025 with respect to non-irradiated value, varying from $n_{Au} = 0.14030 + 3.606\,i$ (non-irradiated value) to $n_{Au} = 0.14055 + 3.606\,i$. Regarding the partial recovery of the SPR spectrum after irradiation, we can reproduce the same behavior considering that the modification of the real refractive index is partially recovered, decreasing around 0.00005 with respect to irradiated value, varying from $n_{Au} = 0.14055 + 3.606\,i$ (irradiated value) to $n_{Au} = 0.14050 + 3.606\,i$. Assuming these variations of Au real refractive index, it is possible to reproduce the slight changes of the measured SPR spectrum in the resonance region (compare experimental results with simulations in Fig. 6.9).

As already mentioned, changes of the Au intrinsic refractive index induced by the X-ray irradiation are not likely. The modification might be rather due to structural or electric modifications at the silica glass/Au interface upon irradiation or it could be also due to the adsorption, induced by irradiation, of molecular or atomic species at the surface of Au, such as atomic oxygen, when the surface is eventually exposed to the ozone [21] generated by synchrotron radiation in the experimental hutch.

These modifications of Au real refractive index for samples with silica substrates involve a slight variation of the order of 0.01 %, which is the same order as the resolution of SPR-XAS setup (see Sect. 5.3 of Chap. 5). For samples with soda-lime glasses, variations of Au index refractive of the order of 1 % were considered, two orders of magnitude larger than for the silica glasses. These differences could be due to the different times of X-ray irradiation (240 min for soda-lime and 170 min for silica substrates) or larger modifications induced at the soda-lime glass/Au interface than at the silica glass/Au interface due to X-ray irradiation.

For samples with silica substrates, the reflectivity for a fixed incident angle close to the resonance (45.6°) was also measured when switching on and off the X-rays in situ and in real time, similar to samples with soda-lime glasses. Figure 6.10a shows the effect of X-ray irradiation on a Au thin film grown onto a silica glass at 11.80 keV. It is possible to observe that the reflectivity increases slightly upon irradiation and after irradiation, it is partially recovered with time. These measurements can be carried out without the Au thin film in reflectivity or transmission mode using the same setup. Therefore, in order to verify that these reflectivity changes are not due to variations on the refractive index of the silica glass substrate, as the fitting of measured SPR spectra indicates, we carried out measurements of transmitted intensity versus time in a silica glass substrate without the Au film, at 11.80 keV. Figure 6.10b shows the result, where we can observe that the intensity variations with the X-rays switched on are the same order than without X-ray irradiation.

Although defects are not observed in the visible range, these can be generated out this range. According to the literature [1, 22, 23], radiation can induce defects on silica glasses associated with either oxygen excess or oxygen defect in the glass matrix. By oxygen excess some defects can be NBOHCs, PORs and interstitial O_3 (ozone) among others, while by oxygen defect some defects can be E' centers

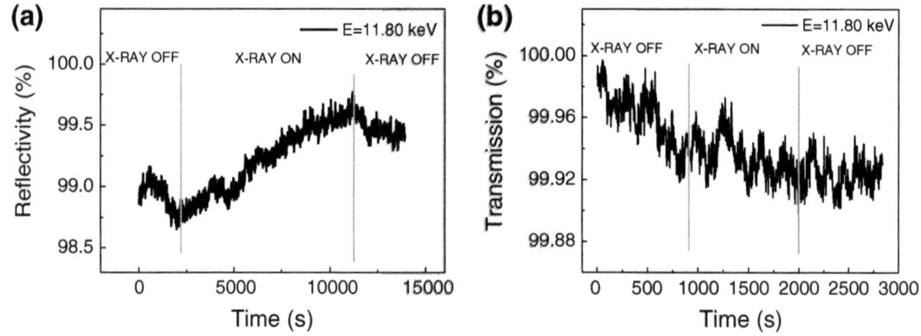

Fig. 6.10 Time dependence of **a** the reflectivity (with a Au film) and **b** the transmission (without the Au film), when switching on and off the X-rays at 11.80 keV, in situ and in real time, for a silica glass

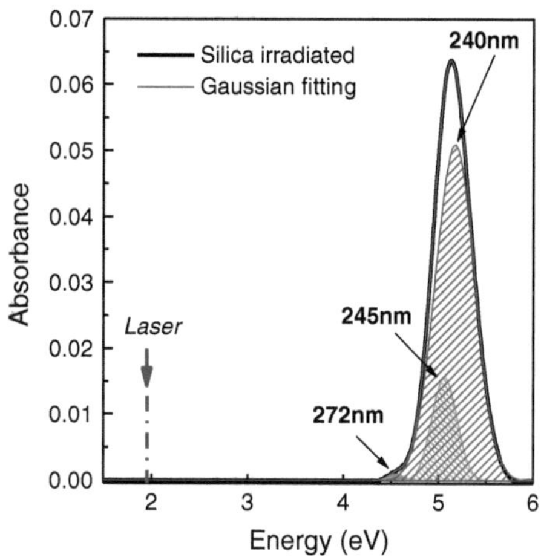

Fig. 6.11 Optical absorption spectrum for a silica glass irradiated at E = 11.80 keV (*black line*) and the fit (*red line*) to three gaussian bands associated with the induced absorptions. The measurement is performed one month after X-ray irradiation. The bands at 240 and 245 nm correspond to B_2 band and the 272 nm band is not identified

or oxygen-deficiency centers: ODCs [1, 22, 31]. Therefore, with the purpose of evaluating the possible defects created in the silica glass when being irradiated with X-rays, optical absorption spectrum of an irradiated silica glass (referenced to non-irradiated silica glass in order to obtain the net irradiation induced absorption) was measured, one month after the irradiation. Figure 6.11 presents the spectrum, which can be decomposed into three gaussian bands (coefficient of determination $R^2 = 0.99997$) that correspond to three absorption bands with maxima around 240, 245 and 272 nm (5.17, 5.06 and 4.55 eV, respectively). Table 6.4 summarizes the main features of these absorption bands.

The observed optical bands [7, 23, 32, 33] can be associated with various defects: NBOHCs (~4.8 eV), PORs (~4.8 eV), interstitial ozone molecule (~4.8 eV), E' centers (~4.8 eV) or ODC defects (~5.0–5.2 eV). Comparing the features of our bands with the reported values [23], the 5.17 and 5.06 eV peaks might be related to ODC defects, which are diamagnetic defects [34]. These optical bands are overlapped showing the named B_2 band [35] in fused silica, which can be decomposed in two types named $B_{2\alpha}$ and $B_{2\beta}$ bands [23, 33], presenting two structurally different ODCs. The $B_{2\alpha}$ band can be attributed to an intrinsic ODC and its structural model can be debated between neutral oxygen vacancy (O≡Si–Si≡O) and twofold coordinated Si (O–$\ddot{\text{S}}$i–O, where ($\ddot{}$) represents two paired electrons in the same orbital) [36, 37].

Table 6.4 Parameters corresponding to the three gaussian fits of the optical absorption bands for silica glass irradiated with X-rays at E = 11.80 keV

Wavelength (nm)	Energy (eV)	FWHM (eV)	Defect
272.5 ± 0.1	4.551 ± 0.002	0.162 ± 0.003	–
244.94 ± 0.02	5.0624 ± 0.0004	0.263 ± 0.001	*ODCs*
239.82 ± 0.02	5.1705 ± 0.0004	0.4113 ± 0.0002	*ODCs*

While some models propose that the $B_{2\beta}$ band corresponds to ODCs perturbed by an impurity [38] or intrinsic ODCs [39]. However, the nature of the optical activity of the B_2 band is still unclear [34]. On the other hand, the 4.55 eV energy peak cannot be identified with any band, but it has been previously observed as permanent damage in high purity fused silica after irradiating [40, 41]. Independently of the defects generated by X-rays in silica glasses, for the previous experiments a 632.8 nm laser was used, hence no absorption band due to defects is detected at that wavelength (see Figs. 6.10b and 6.11).

With these results, we can conclude that for Au thin films grown on silica glass substrates, the X-ray irradiation induces defects in silica glasses but they are not detected in the whole range of the visible spectrum. The changes observed on experimental SPR spectra are small and they are attributed to some kind of modification where the Au thin film is involved, changing its refractive index around 0.01 %. As mentioned above, these variations can be associated with modifications at the silica glass/Au interface upon irradiation or the adsorption of species at the surface of Au film. However, the process is unclear. Due to de robustness of silica glasses under X-ray irradiation, they could be good candidates to be used as substrates, in order to study possible effects of the X-ray irradiation with the SPR-XAS setup in overlayers grown on metallic films.

6.6 Conclusions

The effect of hard X-ray irradiation of soda-lime and silica glasses on their optical properties has been studied. For this purpose, we have used the recently developed experimental setup described in Chap. 5, capable of simultaneously measuring XAS and SPR at a synchrotron beamline. The excitation of surface plasmons in a Au thin film grown on a glass is the probe used to monitor the induced optical changes in the glasses. For soda-lime glasses, two types of defects associated with the observed darkening have been identified: HC1 and HC2 centers. Monitoring, in real time, the reflected intensity versus time, we have determined two different dynamics by recombination with electrons in both the formation and fading of the induced defects. Together with a slow process with a characteristic time of the order of months, we have also identified a fast fading component with a much shorter characteristic time, of the order of minutes and not previously reported. This component has been measured thanks to the setup developed, which allows monitoring changes during and immediately after irradiation. For the case of silica glasses, defects that modify their refractive index have been identified. These non-visible defects are not detected at the laser wavelength used with the SPR-XAS device. However, slight variations have been observed and attributed to changes where the Au thin film is involved.

References

1. E. Friebele, *Radiation Effects in Optical and Properties of Glass* (Westerville, New York, 1991)
2. L.S.G. Pacchioni, D.L. Griscom (eds.), *Defects in SiO_2 and Related Dielectrics: Science and Technology* (Kluwer Academic Publishers, Norwell, 2000)
3. R.A.B. Devine, J.P. Duraud, E. Dooryhee (eds.), *Structure and Imperfections in Amorphous and Crystalline Silicon Dioxide* (Wiley, New York, 2000)
4. H.S. Nalwa (ed.), *Silicon—Based Materials and Devices* (Academic Press, San Diego, 2001)
5. M. Rajaram, E. Friebele, Effects of radiation on the properties of low thermal expansion coefficient materials: a review. J. Non-Cryst. Solids **108**(1), 1–17 (1989)
6. C.E. Nurnberger, R. Livingston, Further studies on the kinetics of the coloring of glass by x-rays. J. Phys. Chem. **41**(5), 691–697 (1937)
7. D.L. Griscom, Optical properties and structure of defects in silica glass. J. Ceram. Soc. Jpn. **99**(1154), 923–942 (1991)
8. J. Sheng, K. Kadono, T. Yazawa, Fading behavior of X-ray induced color centers in soda-lime silicate glass. Applied radiation and isotopes: including data, instrumentation and methods for use in agriculture, industry and medicine **57**(6), 813–817 (2002)
9. J. Sheng, Y. Wu, X. Yang, J. Zhang, UV-laser irradiation on the soda-lime silicate glass. Int. J. Hydrog. Energy **34**(2), 1123–1125 (2009)
10. K. Kadono, N. Itakura, T. Akai, M. Yamashita, T. Yazawa, Effect of additive ions on the optical density and stability of the color centers induced by X-ray irradiation in soda-lime silicate glass. Nucl. Instrum. Methods Phys. Res. Sect. B: Beam Interact. Mater. Atoms **267**(14), 2411–2415 (2009)
11. J. Sheng, K. Kadono, Y. Utagawa, T. Yazawa, X-ray irradiation on the soda-lime container glass. Appl. Radiat. Isot. **56**(4), 621–626 (2002)
12. A. Kameyama, A. Yokotani, K. Kou, Second-order optical nonlinearity and change in refractive index in silica glasses by a combination of thermal poling and x-ray irradiation. J. Appl. Phys. **95**(8), 4000–4006 (2004)
13. N.H. Ky, H. Limberger, R.P. Salathé, F. Cochet, L. Dong, UV-irradiation induced stress and index changes during the growth of type-I and type-II A fiber gratings. Opt. Commun. **225**, 313–318 (2003)
14. W. Knoll, Interfaces and thin films as seen by bound electromagnetic waves. Annu. Rev. Phys. Chem. **49**, 569–638 (1998)
15. A. Serrano, F. Galvez, O. Rodríguez de la Fuente, M.A. Garcia, X-ray irradiation of soda-lime glasses studied in situ with surface plasmon resonance spectroscopy. J. Appl. Phys. **113**(11), 113104 (2013)
16. A. Serrano, O. Rodríguez de la Fuente, V. Collado, J. Rubio-Zuazo, C. Monton, G. Castro, M. García, Simultaneous surface plasmon resonance and x-ray absorption spectroscopy. Rev. Sci. Instrum. **83**(8), 083101 (2012)
17. J.M. Fernández Navarro, *El Vidrio. Consejo Superior de Investigaciones Científicas* (Textos Universitarios, Madrid, 2003)
18. H. Raether, *Surface Plasmons on Smooth and Rouge Surfaces and on Gratings* (Springer, Berlin, 1988)
19. See http://www.mpip-mainz.mpg.de/knoll/soft/
20. L. Dong, J.L. Archambault, L. Reekie, P.S.J. Russell, D.N. Payne, Photoinduced absorption change in germanosilicate preforms: evidence for the color-center model of photosensitivity. Appl. Opt. **34**(18), 3436–3440 (1995)
21. N. Saliba, D.H. Parker, B.E. Koe, Adsorption of oxygen on Au (111) by exposure to ozone. Surf. Sci. **410**(2–3), 270–282 (1998)
22. A. Bishay, Radiation induced color centers in multicomponent glasses. J. Non-Cryst. Solids **3**(1), 54–114 (1970)
23. L. Skuja, M. Hirano, H. Hosono, K. Kajihara, Defects in oxide glasses. Phys. Status Solidi (c) **2**(1), 15–24 (2005)

24. J.W.H. Schreurs, Study of some trapped hole centers in X-irradiated alkali silicate glasses. J. Chem. Phys. **47**(2), 818–830 (1967)
25. D.L. Griscom, Electron spin resonance studies of trapped hole centers in irradiated alkali silicate glasses: a critical comment on current models for HC1 and HC2. J. Non-Cryst. Solids **64**, 229–247 (1984)
26. J. Qiu, X. Jiang, C. Zhu, M. Shirai, J. Si, N. Jiang, K. Hirao, Manipulation of gold nanoparticles inside transparent materials. Angewandte Chemie **43**(17), 2230–2234 (2004)
27. Y. Bensimon, B. Deroide, F. Dijoux, M. Martineau, Nature and thermal stability of paramagnetic defects in natural clay: a study by electron spin resonance. J. Phys. Chem. Solids **61**(10), 1623–1632 (2000)
28. J. Sheng, J. Zhang, L. Qiao, Long-term stability of X-ray induced color centers in silver-doped glass. J. Non-Cryst. Solids **352**(26–27), 2914–2916 (2006)
29. Y. Samoilenko, Y. Kaganovskii, A.A. Lipovskii, E. Mogilko, M. Rosenbluh, CW laser discoloration of X-ray irradiated silver doped silicate glasses. Opt. Mater. **30**(11), 1715–1722 (2008)
30. J. Sheng, X. Yang, W. Dong, J. Zhang, X-ray induced nonbridging oxygen hole center in soda-lime silicate glass. Int. J. Hydrog. Energy **34**(9), 3988–3991 (2009)
31. L. Skuja, Optically active oxygen-deficiency-related centers in amorphous silicon dioxide. J. Non-Cryst. Solids **239**(1–3), 16–48 (1998)
32. K. Awazu, H. Kawazoe, O[sub 2] molecules dissolved in synthetic silica glasses and their photochemical reactions induced by arf excimer laser radiation. J. Appl. Phys. **68**(7), 3584–3591 (1990)
33. N. Kuzuu, X-ray induced absorption bands in synthetic fused silicas: OH content dependence of intensities of X-ray-induced absorption bands in type-III fused silicas. J. Non-Cryst. Solids **179**, 170–178 (1994)
34. M. Cannas, Point Defects in Amorphous SiO_2: Optical Activity in the Visible, UV and Vacuum-UV Spectral Regions. Ph.D. thesis, Università degli studi di Palermo, 1998
35. W.J. Mitchell, J.D. Rigden, The effects of radiation on the near infra-red absorption spectrum of alpha-quartz. Philos. Mag. **2**(20), 941–956 (1957)
36. H. Imai, K. Arai, H. Imagawa, H. Hosono, Y. Abe, Two types of oxygen-deficient centers in synthetic silica glass. Phys. Rev. B **38**, 12772–12775 (1988)
37. L. Skuja, A. Streletsky, A. Pakovich, A new intrinsic defect in amorphous SiO_2: twofold coordinated silicon. Solid State Commun. **50**(12), 1069–1072 (1984)
38. A. Anedda, F. Congiu, F. Raga, A. Corazza, M. Martini, G. Spinolo, A. Vedda, Time resolved photoluminescence of a centers in neutron irradiated SiO_2. Nucl. Instrum. Methods Phys. Res. Sect. B: Beam Interact Mater. Atoms **91**, 405–409 (1994)
39. N. Kuzuu, M. Murahara, Excimer-laser-induced emission bands in fused quartz. Phys. Rev. B 3083–3088 (1993)
40. E. Friebele, D. Griscom, G. Siger, *The Physics of Non-Cryst Solids* (Trans Tech Publications, Aedermansdorf, 1977)
41. D. Griscom, Point defects and radiation damage processes in alpha-quartz, in *33rd Annual Symposium on Frequency Control*, 1979

Chapter 7
Study of Co-Phthalocyanines Irradiated with X-Rays by Surface Plasmon Resonance

This chapter presents a study of the effects of hard X-ray irradiation on Co-phthalocyanine thin films using the device described in Chap. 5. Co-phthalocyanine films with different thickness and growth temperature are studied. The very high sensitivity of surface plasmon resonance to slight modifications of the dielectric media allows tracking the small changes of refractive index induced by the irradiation in the organic films.

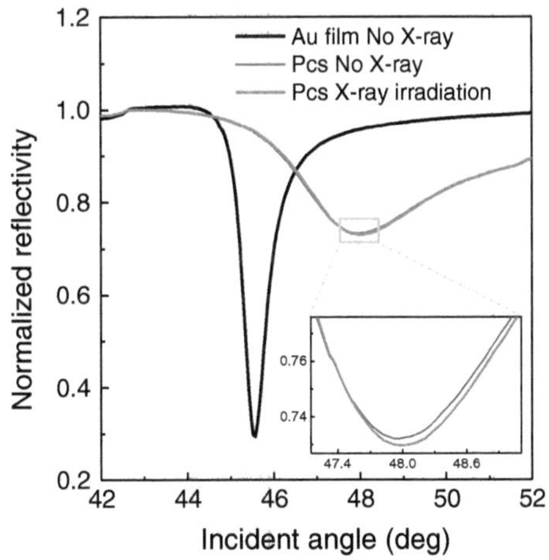

7.1 Introduction

Phthalocyanines (Pcs) represent an important type of organic semiconductor showing optical and electronic properties [1, 2] with a wide range of applications such as in electrochromic devices, field effect transistors, light emitting diodes [3], gas sensors [4, 5] or as dye molecules for solar cells [6]. Pcs ($C_{32}H_{16}N_8{}^{-2}$) are polyaromatic molecules that can accommodate a large variety of metal ions

A. Serrano Rubio, *Modified Au-Based Nanomaterials Studied by Surface Plasmon Resonance Spectroscopy*, Springer Theses, DOI 10.1007/978-3-319-19402-8_7

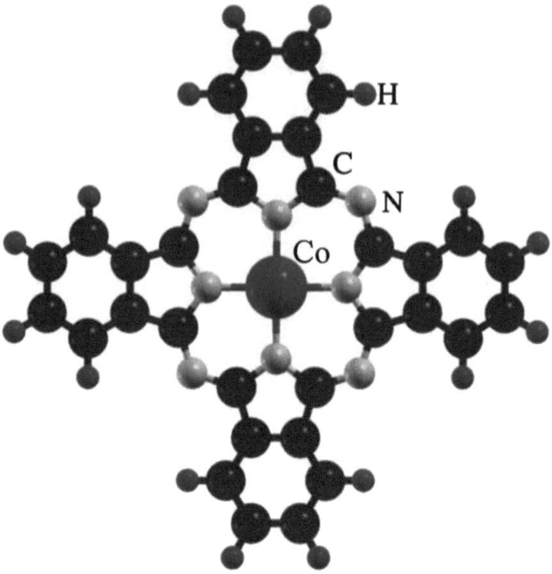

Fig. 7.1 Structure of a CoPc molecule, showing the four-leaf-clover structure with the Co metal at the center. Figure taken from Bhattacharjee et al. [10]

(i.e., Co^{2+}, Cu^{2+}, Mn^{2+}, Zn^{2+}, Fe^{2+}) in the central cavity [7], as shown Fig. 7.1, where a Co metal is found at the molecule center. The central cavity of the Pcs can also be empty and then they are called metal-free phthalocyanine (H_2Pc) [8]. Specifically, Co-phthalocyanines (CoPcs) are already applied in optoelectronic devices, gas sensors and they are considered a promising material for the development of low-dimensional organic molecular magnets [9].

Pcs are thermally and chemically stable [11] and their versatility plays an important role in material science and nanotechnology [7]. Thus, Pcs are usually studied and characterized by synchrotron radiation for probing different processes that occur at the atomic or the molecular level, as for example the charge transfer at the interface between a CoPc monolayer and Au [9]. However, an important obstacle of these investigations delays in the radiation damage that can affect the electronic structure of molecules, making more difficult to achieve studies of these beam-sensitive materials. Some damages induced by radiation can be the break of the peptide bond between amino acids of a protein, a change in the 3d electron configuration of the transition metal center in a protein or a shift in the Fermi level due to the creation of defect states in the band gap of dye molecules [12]. In the literature, some studies can be found about the radiation damage on Pcs [12, 13]. Cabailh et al. observed changes in the intensity signal of Sn-phthalocyanines grown on GaAs(001) as a function of exposure to the radiation at nitrogen 1 s edge, but they did not find a clear explanation [13]. Later, Cook et al. found that the radiation on Pcs with soft X-rays affects the electronic structure of the molecules inducing a shift in the Fermi level between the HOMO (highest occupied molecular orbital) and LUMO (lowest unoccupied molecular orbital) due to radiation-induced gap states in the Pcs [12]. This modification may be visible in their optical properties [4, 12, 14].

It is well known that the morphological properties of Pc thin films depend on parameters such as their thickness and growth conditions, which control the degree of crystallinity, grain size or the crystallographic phase. The deposition of Pc films depends, among other factors, on the substrate temperature during deposition [15, 16]. Usually, films deposited on amorphous substrates at room temperature (RT) exhibit the so-called α-phase while deposition temperatures above 150 °C lead to films with β-phase, which is characterized by elongated crystallites, with a larger size [15, 17] and a more compact structure [7] than that of the films grown at RT. These phases exhibit different optical properties [18], thus the induced variations by X-rays on Pcs could be different depending on the Pc thickness or the crystallinity [12, 14].

Surface plasmon resonance (SPR) is high sensitivity to any modification of the dielectric media close to the metallic film. Thus, changes on organic molecules deposited over the metallic film can be analyzed, as shown in Chap. 5, providing simultaneously information on the variations of both real and imaginary parts of their refractive index [19]. In this work, we present a study of SPR on CoPc/Au samples upon X-ray irradiation using the SPR-XAS setup [20] described in Chap. 5. Here, we study in situ and real time the effects of X-ray irradiation on the optical properties of CoPcs by SPR [21], changing the CoPc film thickness and growth temperature.

7.2 Fabrication of Samples

CoPc/Au samples were grown on silica substrates in two steps, using the home-made evaporation chamber of the Schuller Nanoscience Group, at the University of California, San Diego. First, Au thin films with nominal thicknesses around 60 nm were fabricated by thermal evaporation at RT under a 10^{-6} torr deposition pressure. Subsequently, CoPc thin films were thermally evaporated, as explained in Sect. 2.2.1 of Chap. 2, covering only half of the samples, as shown Fig. 7.2. The area of the silica substrate was of 2×2 cm^2. Samples with CoPc thin films with nominal thicknesses of 2, 5 and 10 nm were prepared at RT and at 200 °C. Table 7.1 shows a description of samples fabricated. Deposited Au and CoPc thicknesses were controlled by a Q-microbalance.

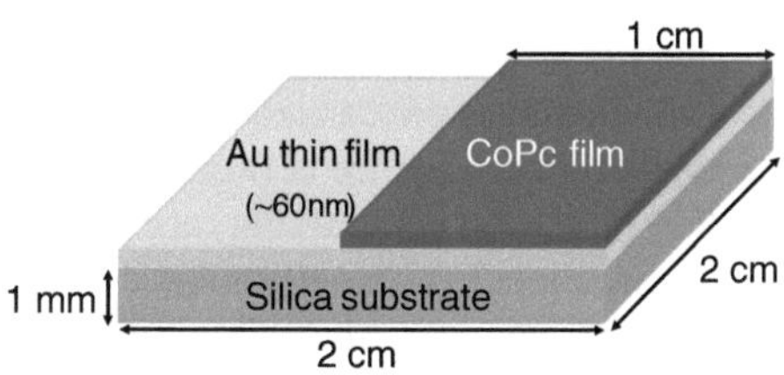

Fig. 7.2 Configuration of CoPc/Au samples grown on silica substrates

Table 7.1 CoPc/Au samples with CoPc thin films of nominal thicknesses of 2, 5 and 10 nm, grown at RT and at 200 °C on silica substrates

Sample	Description
2-RT	2 nm CoPc layer grown at RT on 60 nm Au/silica substrate system
5-RT	5 nm CoPc layer grown at RT on 60 nm Au/silica substrate system
10-RT	10 nm CoPc layer grown at RT on 60 nm Au/silica substrate system
2–200	2 nm CoPc layer grown at 200 °C on 60 nm Au/silica substrate system
5–200	5 nm CoPc layer grown at 200 °C on 60 nm Au/silica substrate system
10–200	10 nm CoPc layer grown at 200 °C on 60 nm Au/silica substrate system

7.3 Experimental Details

Experiments were carried out at the branch A of the BM25 SpLine beamline at the European Synchrotron Radiation Facility (ESRF) in Grenoble (France), by the SPR-XAS device [20] previously described in Chap. 5. The followed protocol for the measurements was similar to that performed in Chap. 6. Experimental SPR spectra were measured before, during and after X-ray irradiation around the Co K edge: at 7.72 and 7.75 keV that correspond to energies at and above the absorption edge, respectively. For all samples, the X-ray irradiation was carried out around 60 min.

More interesting information about the effects of the X-ray irradiation on CoPcs might be obtained studying the SPR spectra when irradiating around the N or C absorption edge (hundreds of eV). For that, we would have to mount the setup at a soft X-ray beamline and the experiments would have to carry out in vacuum, since the air strongly absorbs X-rays with energy below 5 keV. Our SPR-XAS device [20] is designed to carry out experiments in air, therefore in this work we analyze the samples irradiated with hard X-rays instead of with soft ones and so, at the Co absorption edge.

Simulations of SPR spectra were performed with the freeware software Winspall [22]. Known the effects on the Au/substrate systems (see Chap. 6), iterative simulations were carried out for CoPc/Au samples. Fits were taken good when the simulation divergence between the measured and simulated SPR spectra was the smallest. The Raman scattering spectra of CoPc layers were investigated on different regions irradiated and non-irradiated with X-rays in order to analyze any possible structural damage on CoPcs. The final spectra are an average of 40 spectra integrating 5 s each spectrum. Raman measurements were performed 2 months after irradiating the samples.

7.4 Study of Co-Phthalocyanines by SPR-XAS Setup

In Chap. 6, we studied the effects induced by X-ray irradiation in glass substrates. We found important modifications of the refractive index for soda-lime substrates upon X-ray irradiation. On the contrary, for silica glasses, the effects of the X-ray irradiation were very small and close to the detection limit of the SPR-XAS device. Hence, in order to study the possible modifications generated by X-rays on CoPcs, CoPc/Au samples were grown on silica substrates instead of soda-lime substrates.

In order to discriminate between the effects of X-ray irradiation on the CoPcs and those on the Au film grown and silica substrate, both regions of all samples (CoPc/Au/silica and Au/silica) were irradiated with X-rays under same conditions of irradiation time, energy and intensity.

7.4.1 SPR of Au Thin Films Irradiated with X-Rays

Figure 7.3 shows experimental SPR spectra for the 60 nm bare Au film deposited onto a silica substrate of 5-RT sample before, during and after X-ray irradiation at 7.72 keV (Co K edge). No significant variations are found in the reflected intensity upon X-ray irradiation for 60 min, similar to results obtained in Chap. 6 for Au films grown onto silica substrates irradiated for 170 min at Au L3 edge. A detail of the resonance region reveals that X-ray irradiation induces slight changes on SPR spectra, decreasing the reflected intensity around 0.003 and widening the resonance region about 0.01° at the full width half maximum (FWHM). Moreover, we observe

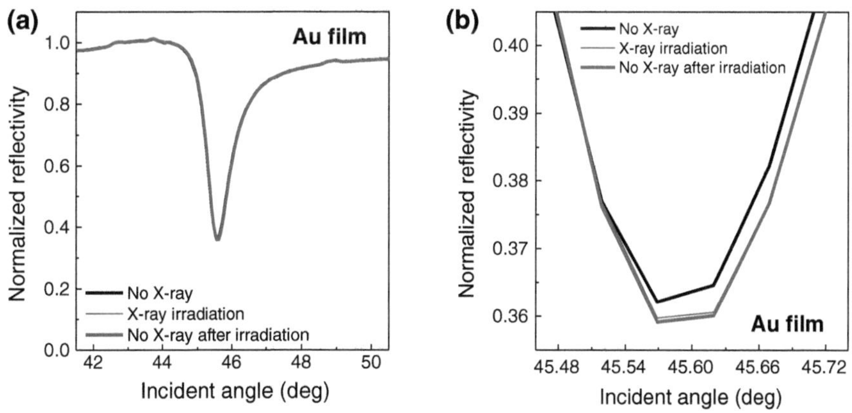

Fig. 7.3 **a** SPR curves for a 60 nm bare Au film on silica glass before (*black line*), during (after 60 min of X-rays, *blue line*) and after (after 40 min without X-rays, *red line*) irradiation with X-rays at 7.72 keV. **b** A detail of the resonance region. Spectra were normalized to critical angle

that these variations are mostly irreversible and after switching off the X-rays, after 40 min without X-ray irradiation, the effects remain present.

In order to estimate the variations induced by X-ray irradiation in the bare Au deposited on a silica substrate, SPR simulations were performed. The SPR spectrum of Au film grown onto silica glass prior to the X-ray irradiation was fitted with a thickness of 61 nm for the Au film and using $n_{prism} = 1.457$ and $n_{Au} = 0.1957 + 3.8106\,i$ at $\lambda = 633$ nm. Afterwards, simulation of the measured SPR when Au film is irradiated with X-rays was performed, varying the refractive index of the modified layer, in order to reproduce the experimental results presented in Fig. 7.3. In this fit, we obtain a very good agreement with the experimental results assuming that the refractive index of the modified layer (n_{Au}) upon irradiation changes with respect to non-irradiated value, varying from $n_{Au} = 0.1957 + 3.8106\,i$ (non-irradiated value) to $n_{Au} = 0.1956 + 3.8112\,i$ (irradiated value). For that, the Au film is considered as the modified medium so its refractive index is the only free parameter, while the rest of parameters (included those of the silica glass) remain fixed with the same values as those prior to X-ray irradiation. Figure 7.4 shows the results of simulated SPR curves and a detail of resonance region compared with the resonance region of measured SPR spectra. These modifications of Au refractive index mean a slight variation of the order of 0.01 %, which are the same order than those obtained in

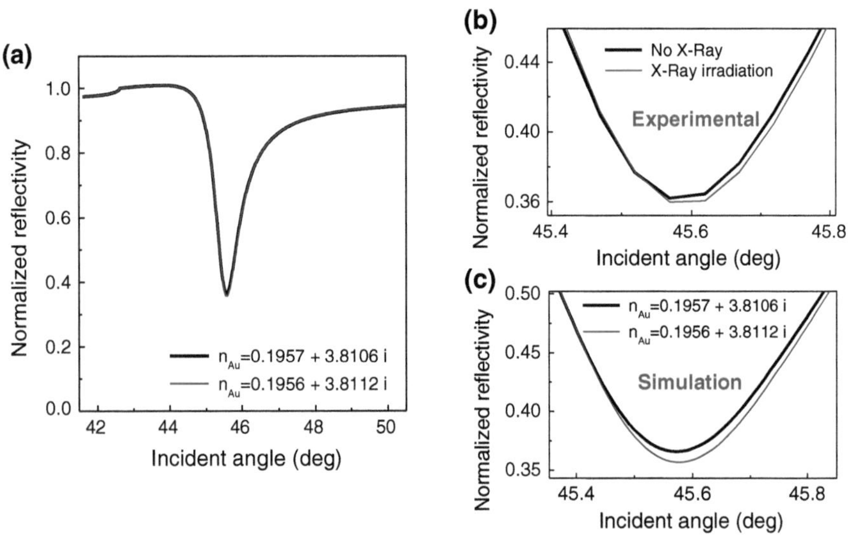

Fig. 7.4 **a** Simulated SPR spectrum using the parameters indicated in the legend for a 61 nm Au film onto silica glass substrates before (*black line*) and during (*blue line*) irradiation with X-rays at 7.72 keV. **b** A detail of the resonance region for the normalized experimental spectra and **c** a detail of the resonance region for the normalized calculated spectra

Chap. 6 for the samples with silica substrates irradiated at the Au L3 absorption edge. Possible differences of refractive index variations could be found due to the different conditions of X-ray irradiation or different concentration of ozone into the experimental hutch [23].

7.4.2 SPR of CoPc/Au Samples Irradiated with X-Rays

Taking into account the slight variations of the Au refractive index induced by X-ray irradiation for the bare Au region deposited on silica substrate, of the order of 0.01 %, we measured the SPR on the region of the CoPc/Au bilayers on silica substrates varying the thickness of CoPc layer and the growth temperature, in order to study the effects of X-rays on the CoPc layers.

Figures 7.5, 7.6 and 7.7 show the experimental SPR spectra related to CoPcs with thicknesses 2, 5 and 10 nm respectively, grown at RT and 200 °C on a Au film. For samples irradiated with X-rays for 60 min, modifications in the resonance region of the curves are observed, which depend on CoPc thickness and growth temperature. As shown below, these variations are larger than those observed in the bare Au region and are weakly recovered after X-ray irradiation. For sample 2-RT (see Fig. 7.5a, b) irradiated with X-rays, a widening of the SPR curve is observed with an increase of the FWHM around 0.03°, a shift at the resonance region towards larger angles of 0.01° and a decrease of the intensity about 0.01. For CoPcs grown at 200 °C, (sample 2–200, see Fig. 7.5c, d), with the X-ray irradiation, the resonance angle is shifted towards lower angles about 0.02°, the intensity decreases around 0.006 and the SPR spectrum presents a widening at the FWHM of 0.02°. For the case of sample 5-RT (see Fig. 7.6a, b) irradiated with X-rays, the SPR curve shows a widening at the FWHM around 0.06°, a shift at the resonance angle towards larger angles of 0.03° and a decrease of the intensity about 0.01. For sample 5–200 (see Fig. 7.6c, d), with the X-rays, the resonance region is shifted towards lower angles about 0.04°, the intensity increases around 0.004 and the SPR spectrum shows a widening at the FWHM of 0.03°. For sample 10-RT (see Fig. 7.7a, b) after X-ray irradiation, it is possible to observe that the SPR curve is modified widening at the FWHM around 0.08°, the resonance region is shifted towards larger angles around 0.03° and the intensity decreases about 0.03. Finally, for sample 10–200 (see Fig. 7.7c, d) irradiated with X-rays, the resonance angle of SPR curve is shifted towards larger angles about 0.06°, the intensity decreases around 0.006 and the SPR spectrum presents a widening at the FWHM of 0.03°. As summary, Table 7.2 shows the values of the changes observed with the X-rays on SPR spectra for CoPcs grown on Au film/silica substrate systems and for the bare Au region. For all cases, SPR variations for CoPc/Au bilayers are larger than those observed on the bare Au regions. Therefore, the X-ray irradiation induces changes on SPR spectra of CoPc/Au bilayers due to modifications generated in CoPcs. Variations of SPR reflectivity and the FWHM are more accused in the samples grown at RT than in those grown at 200 °C. Moreover, the FWHM changes in the resonance region depend on CoPc layer thickness, for both CoPcs grown at

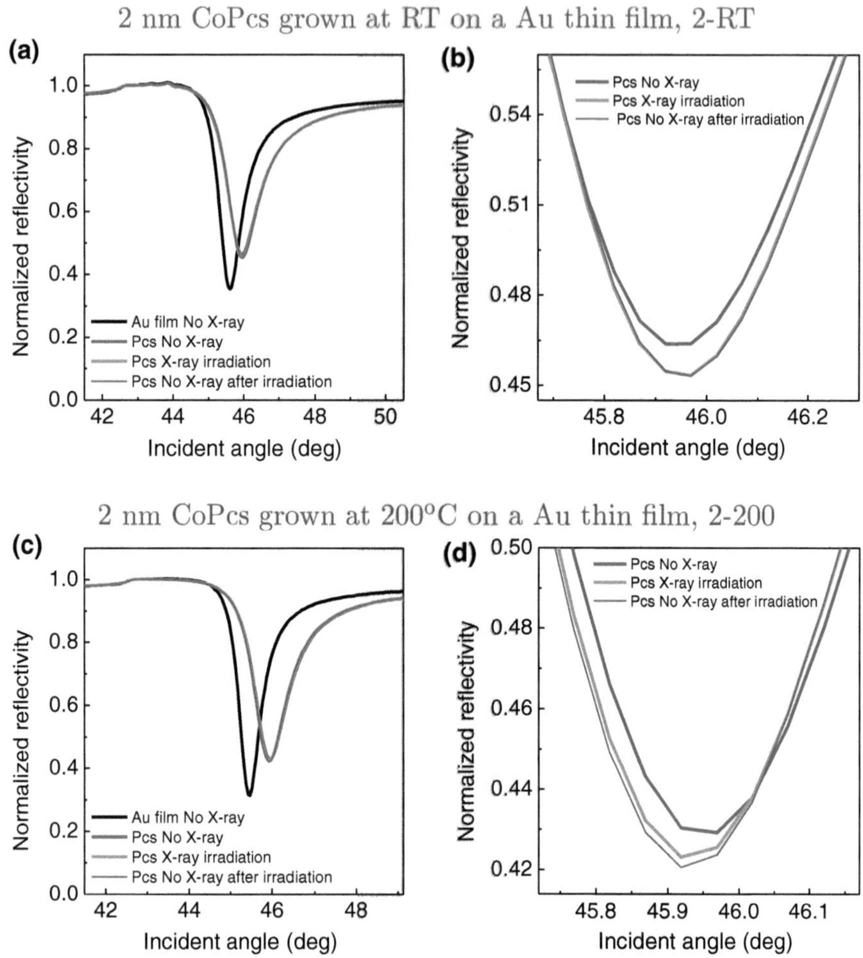

Fig. 7.5 SPR curves for the bare Au film (*black line*) and the 2 nm CoPc/Au bilayer on silica glass before (*red line*), during (after 60 min with X- rays, *green line*) and after (after 40 min without X-rays, *blue line*) irradiation with X-rays at 7.72 keV. **a** and **c** CoPc layers grown at RT and 200 °C respectively and **b** and **d** a detail of the resonance region. Spectra were normalized at the critical angle

RT and at 200 °C, observing larger modifications when the layer thickness increases. Modifications of the θ_R do not show a clear tendency, neither with CoPc thickness nor with growth temperature (see Table 7.2).

In general, after 40 min without X-ray irradiation, neither the shift nor the widening of the resonance are reversible after irradiation. Moreover, as expected, the critical angle does not change upon X-ray irradiation since it depends just on the refractive index of the quartz prism and the air (see Sect. 2.10 of Chap. 2).

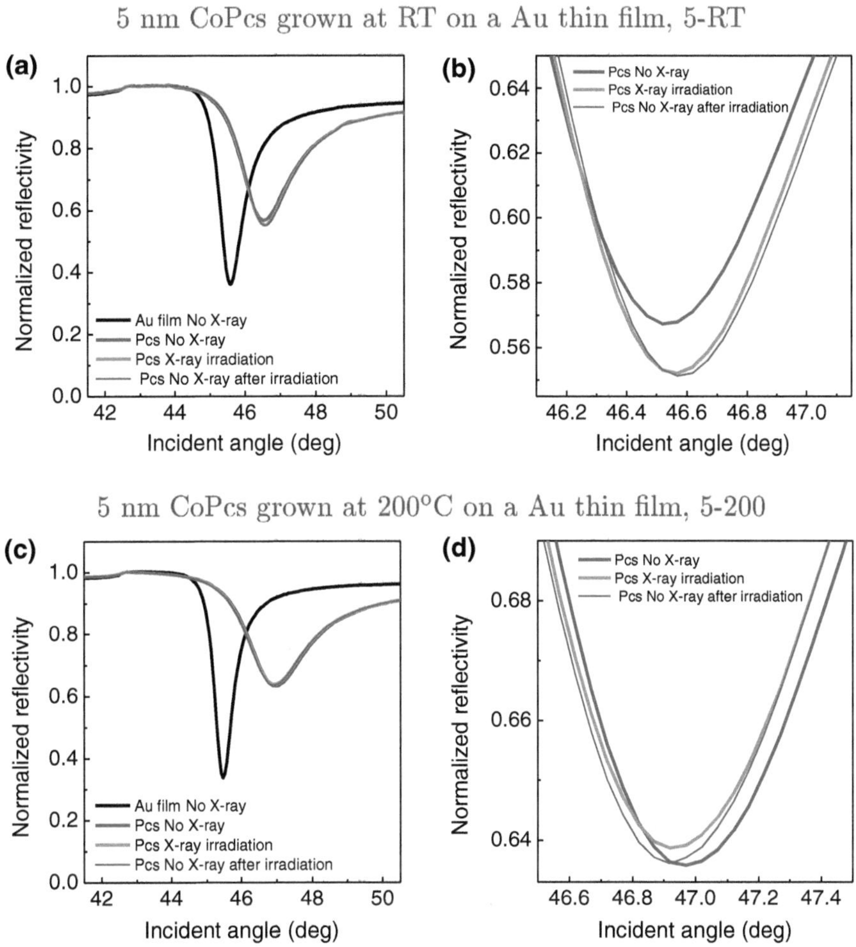

Fig. 7.6 SPR curves for the bare Au film (*black line*) and the 5 nm CoPc/Au bilayer on silica glass before (*red line*), during (after 60 min with X- rays, *green line*) and after (after 40 min without X-rays, *blue line*) irradiation with X-rays at 7.72 keV. **a** and **c** CoPc layers grown at RT and 200 °C respectively and **b** and **d** a detail of the resonance region. Spectra were normalized at the critical angle

In order to estimate the changes induced on the refractive index of CoPc layers when CoPc/Au samples are irradiated with X-rays, we simulated the SPR curves before and during (after 60 min) X-ray irradiation. First, we fitted the SPR spectra for each bare Au region of each sample before X-ray irradiation. For each sample, we found the best values of refractive index and thickness for the Au film, considering a four-media system. Subsequently, we performed the simulations of the SPR spectra of CoPc/Au bilayers on silica substrates without irradiation. For that, we used the calculated parameters of each bare Au region for each sample and we varied the refractive index and the thickness of the CoPc layers, in our five-media system

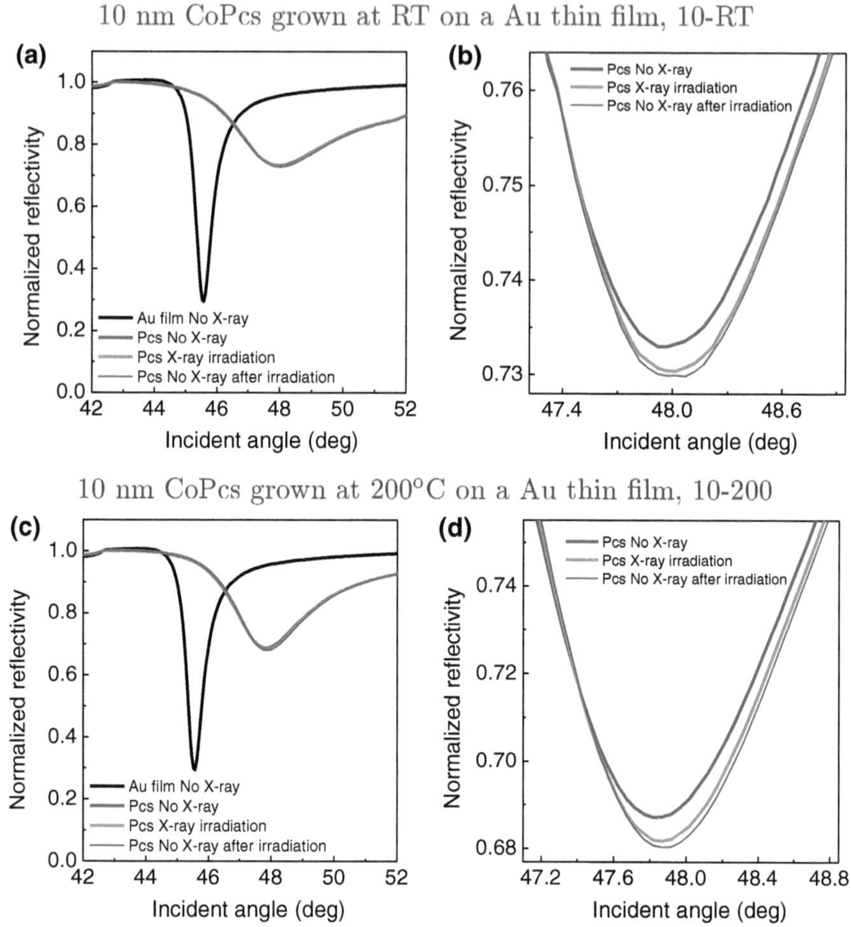

Fig. 7.7 SPR curves for the bare Au film (*black line*) and the 10 nm CoPc/Au bilayer on silica glass before (*red line*), during (after 60 min with X- rays, *green line*) and after (after 40 min without X-rays, *blue line*) irradiation with X-rays at 7.72 keV. **a** and **c** CoPc layers grown at RT and 200 °C respectively and **b** and **d** a detail of the resonance region. Spectra were normalized at the critical angle

(last four-media system + CoPc layer), in order to reproduce the experimental results before the irradiation. Once calculated the parameters of CoPc layers before the X-ray irradiation, we fitted the SPR curves during irradiation. For that, we assumed that the CoPc layer thicknesses are not modified by X-rays and that the refractive index is the only free parameter. The rest of parameters remain fixed with the same value as those prior to X-ray irradiation. For all cases, we obtain a very good agreement with the experimental results.

Figures 7.8 and 7.9 show the SPR simulations corresponding to samples 2–200 and 10-RT, respectively. For sample 2–200, we obtain that the refractive index of Au film is $n_{Au} = 0.1583+3.8109\,i$ and its thickness is 61.7 nm and for the modified layer

Table 7.2 Modifications in the resonance region of SPR spectra of CoPcs grown on Au/silica samples and of the bare Au film after X-ray irradiation at 7.72 keV, for 60 min

X-ray effects	Bare Au film	2-RT	2–200	5-RT	5–200	10-RT	10–200
Changes at FWHM (°)	+0.01	+0.03	+0.02	+0.06	+0.03	+0.08	+0.06
Shift of θ_R (°)	–	+0.01	−0.02	+0.03	−0.04	+0.03	+0.03
Intensity changes	+0.003	+0.01	+0.006	+0.01	−0.004	+0.03	+0.006

The +/− symbol represents SPR changes due to the irradiation towards larger/slower values with respect to the SPR spectrum of samples without irradiating

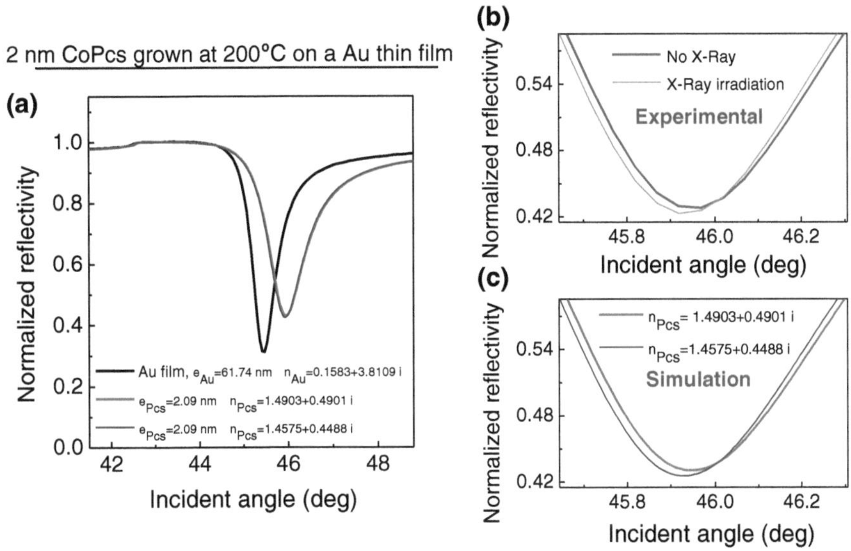

Fig. 7.8 **a** Simulated SPR spectra using the parameters indicated in the legend for sample 2–200, assuming a 61.7 nm Au film thickness and a thickness of 2.1 nm for the CoPc layer, before (*pink line*) and during (*purple line*) irradiation with X-rays at 7.72 keV. **b** A detail of the resonance region for the normalized experimental spectra and **c** a detail of the resonance region for the normalized calculated spectra

(CoPc layer) upon irradiation, with a simulated thickness of 2.1 nm, the refractive index (n_{CoPc}) changes from $n_{CoPc} = 1.4903 + 0.4901\,i$ (non-irradiated value) to $n_{CoPc} = 1.4575 + 0.4488\,i$ (irradiated value). For sample 10-RT, the refractive index of Au film is $n_{Au} = 0.1584 + 3.8248\,i$ and the simulated thickness is of 60.2 nm while for the CoPc layer upon irradiation, with a simulated thickness of 8.2 nm, the refractive index (n_{CoPc}) changes from $n_{CoPc} = 1.8882 + 0.9008\,i$ (non-irradiated value) to $n_{CoPc} = 1.9047 + 0.8858\,i$ (irradiated value).

These modifications of the refractive index of CoPc layers with X-rays mean variations around 1.5 % for sample 10-RT and around 4 % for sample 2–200. Simulations of SPR spectra for rest of samples showed variations of the refractive index of

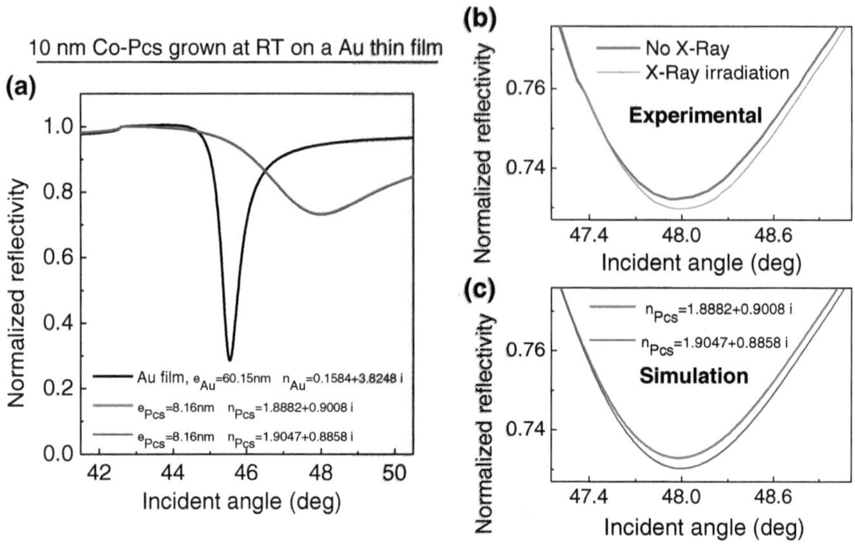

Fig. 7.9 **a** Simulated SPR spectra using the parameters indicated in the legend for sample 10-RT, assuming a 60.2 nm Au film thickness and a thickness of 8.5 nm for the CoPc layer, before (*pink line*) and during (*purple line*) irradiation with X-rays at 7.72 keV. **b** A detail of the resonance region for the normalized experimental spectra and **c** a detail of the resonance region for the normalized calculated spectra

CoPc layers between 1.5 and 5 %, which change depending on the different thickness and growth temperature of CoPc layers. These modifications are about two orders of magnitude larger than those observed in the bare Au regions (0.01 %) irradiated under the same conditions than the CoPc/Au bilayers. Thus, we conclude that X-ray irradiation on CoPc/Au bilayers induces modifications in the CoPc layers, which modify both real and imaginary parts of refractive index.[1]

Using the SPR-XAS device (described in Chap. 5), the reflectivity was measured for a fixed incident angle close to the resonance (45.95°), when switching on and off the X-rays. Figure 7.10 shows the effects of X-ray irradiation at 7.75 keV, above the Co K edge, for the sample 2-RT. The results confirm that the reflectivity changes, decreasing, upon irradiation with X-rays and after irradiating, the reflectivity is slowly recovered. Most of these modifications can be ascribed to the organic films, since the reflectivity changes at the resonance angle for the Au/silica zone (without CoPcs) are around an order of magnitude smaller than for the CoPcs grown on Au/silica (see Table 7.2). Furthermore, we measured with the sample placed at an incident angle of 45.95°, region away from the resonance region of the SPR spectrum for the Au/silica zone and where the reflectivity changes are around 5 % smaller than those close to the

[1] Simulations of SPR results after X-ray irradiation were performed leaving free the thickness besides the refractive index of the CoPc layers, however we only obtained a good agreement for thicker CoPc layers, somewhat inconsistent with the experiment.

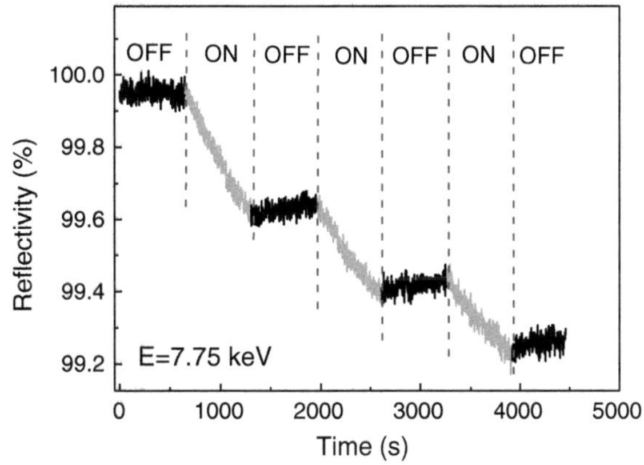

Fig. 7.10 Reflected intensity versus time when switching on and off the X-rays at 7.75 keV, in situ and in real time for 2 nm CoPcs grown on Au film/ silica glass at RT

resonance angle. Therefore, the reflectivity variations that could be associated with modifications of the Au/silica zone are small and can be neglected. The measured modifications of the reflectivity versus time, for the sample 2-RT, can be associated with changes induced in the CoPcs.

Therefore, we can conclude that CoPc/Au bilayers irradiated with X-rays show SPR spectra modified with respect to samples without irradiating. Studying bare Au regions under the same conditions than CoPc/Au bilayers, we could discard that the measured effects induced by X-rays on CoPc/Au bilayers are due to modifications in Au films grown on silica substrates. X-ray irradiation induces changes on CoPcs modifying their refractive index, around 1.5–5 %, depending on the CoPc film thickness and growth temperature. The modifications of SPR spectra for CoPc/Au bilayers caused by X-ray irradiation, can be related to the different features of the molecules depending on the thickness and growth conditions. Variations of SPR reflectivity and the FWHM are smaller for CoPc films grown at 200 °C, which present a β-phase and show a more compact structure [7] than CoPcs grown at RT that exhibit the α-phase. Besides, independently of the growth temperature, thicker CoPc films show larger FWHM modifications of SPR spectra. Thus, the changes in optical properties of CoPc films irradiated with X-rays depend on their thickness and growth temperature.

Experiments were performed at the beamline hutch, in air. In order to discard that the observed effects on SPR spectra are due to an instable conductivity in CoPcs and/or the adsorption of molecular or atomic species on the surface of CoPc layers, measurements of the reflectivity versus time were carried out in He atmosphere, in order to stabilize the conductivity of CoPcs and avoid possible adsorption on the CoPc surface. In He atmosphere, we observed the same effects than in air, therefore the modifications induced by X-rays on the SPR spectra are associated with changes in the physical properties of CoPcs.

7.5 Raman Spectroscopy of CoPc/Au Samples Irradiated with X-Rays

In order to analyze if the changes induced with X-ray irradiation on the refractive index of CoPc layers are due to structural modifications, samples were studied by Raman spectroscopy. This technique was chosen since can provide information about the order, crystal size or defects of samples [24], as explained in Chap. 2. For all samples, pristine regions and regions where CoPc/Au bilayers were irradiated for 60 min at 7.72 keV were studied. Four scans on the irradiated regions and two spectra outside the irradiated regions were measured on the same day and analysis conditions, in order to avoid possible errors in the measurements due to aging effects of the CoPcs with time [21, 25].

The vibrations of the CoPc planar molecule (a scheme is shown in Fig. 7.1), which consists of 57 atoms, can be classified into the following irreducible representation (taking into account only internal vibrations) [26–29]:

$$\Gamma_{vibrational} = 14A_{1g} + 13A_{2g} + 14B_{1g} + 14B_{2g} + 13E_g + 6A_{1u} + 8A_{2u} + 7B_{1u} + 7B_{2u} + 28E_u \quad (7.1)$$

These modes can be divided into three categories: modes active in Raman scattering, modes active in IR spectra and modes that are neither active in Raman nor IR spectra (but these become active in Raman, as a result of an interaction with the support, if a distortion of the molecular symmetry occurs). The basic Raman selection rules allow A_{1g}, B_{1g}, B_{2g}, and E_g modes in the normal Raman scattering. The non degenerate A_{1g}, B_{1g} and B_{2g} modes are in-plane vibrations, and double degenerate E_g mode is the out-of-plane vibration. The A_{2g} modes occur in the resonance spectrum as a result of the vibronic coupling between two electronic excited states of Pcs [30]. Thus, 68 Raman modes can become activated. Raman spectra can also exhibit overtones and combination bands, in addition to the fundamentals. These can be obtained in the region 1800–3800 cm^{-1}.

Figures 7.11, 7.12 and 7.13 show the Raman spectra for CoPc/Au bilayers with CoPc thicknesses: 2, 5 and 10 nm respectively, presenting two spectra for the irradiated region with X-rays and one spectrum outside the irradiated region. The rest of measured spectra were similar.

For samples with CoPc layers with thicknesses of 2 and 5 nm grown at RT and 200 °C, the Raman spectra in the irradiated regions result more intense than in the pristine zone (see Figs. 7.11 and 7.12). However, for samples 10-RT and 10–200, we observe the opposite behavior: Raman signal is larger in the pristine zone than in the irradiated one (see Fig. 7.13). For all samples, neither a shift nor other difference for the Raman modes are apparently observed between the irradiated and non-irradiated regions, as shown a detail of the Raman spectra, from 550 to 1600 cm^{-1} (see Figs. 7.11b, d, 7.12b, d, 7.13b, d). There is not a clear correlation between the changes observed in the Raman spectra and those observed in SPR ones.

The differences in the intensity of the Raman signal between spectra collected in the same sample (Figs. 7.11, 7.12 and 7.13) could be ascribed to different origins:

1. Different emission intensities related to changes in the material concentration [24]. A degradation of the CoPc upon irradiation yielding a decrease in the amount of CoPc could modify the intensity of the Raman signal. On the one hand, for this thickness range, the Raman signal is about proportional to the thickness of the film. On the other hand, a reduction in the amount of CoPc can increase Au fluorescence and promote surface enhanced Raman scattering (SERS) [31], increasing the Raman signal. However, the changes in the intensity detected in the spectra are about the same for the background (related to Au fluorescence) than for the Raman bands (associated with CoPc layer), so we can discard significant material removal upon X-ray irradiation. Moreover, we tried to simulate the SPR spectra of CoPc/Au bilayers measured during the irradiation with X-rays (Figs. 7.8 and 7.9) leaving free the thickness of CoPc layers besides the refractive index. We only obtained a good agreement for thicker CoPc layers, somewhat inconsistent with the experiment.
2. Differences related to local crystal texturation. As earlier described in Chap. 4, the intensity of each Raman band depends on the relative orientation of the crystal and the electric field of the light. This effect makes the relative intensity

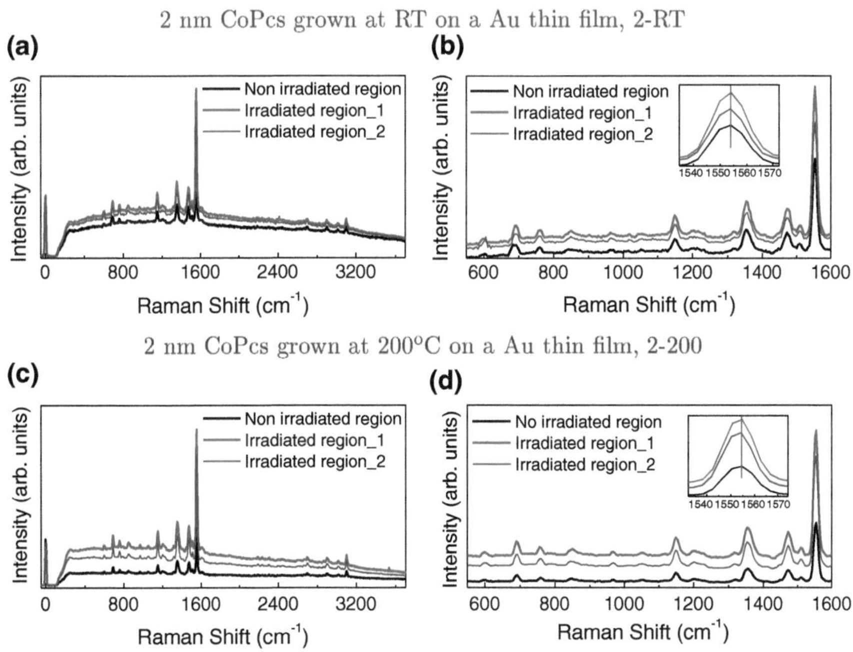

Fig. 7.11 Raman spectra on irradiated and non-irradiated regions for **a** 2-RT **c** 2–200. **b** and **d** a detail from 550 to 1600 cm^{-1} of Raman spectra

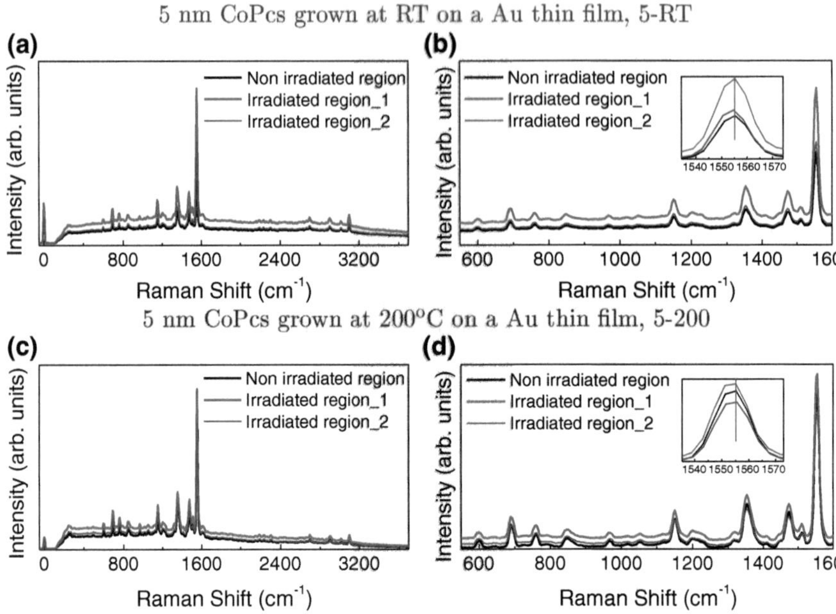

Fig. 7.12 Raman spectra on irradiated and non-irradiated regions for **a** 5-RT and **c** 5–200. **b** and **d** a detail from 550 to 1600 cm^{-1} of Raman spectra

of the Raman bands can change when having different orientations. We observe here that the bands relative intensity is about constant, discarding texturation effects of CoPc layers.

3. Changes in the experimental conditions. Optical measurements collecting light emission are very sensitive to relative position of the sample and the optical detectors. This sensitivity is extremely high in the case of confocal measurements for thin films, where the collected signal strongly changes with a small variation of the sample-optical detector distances. In our measurements, when moving from one region to another in order to collect the different spectra, it is possible that the sample-optical detector distances slightly changes because of a small misalignment due to slight tilt of sample. These experimental modifications should affect equally to the signals coming from Au and CoPc film, as it is observed in our measurements.
4. Inhomogeneity in the thickness or local structural features of CoPc initial layers, which should lead to different intensities in the Raman signals. In this case, the differences observed between two spectra collected at irradiated regions should be of the same order of magnitude than those observed between spectra from an irradiated and non-irradiated regions. This is what we experimentally observe, as shown Table 7.3 where four vibrational modes of the Raman spectra from irradiated regions and without irradiating of samples 2-RT and 10–200 are fitted to lorentzian bands. On the other hand, the homogeneity of CoPc layers can be evaluated by confocal Raman microscopy on a non-irradiated area. For sample

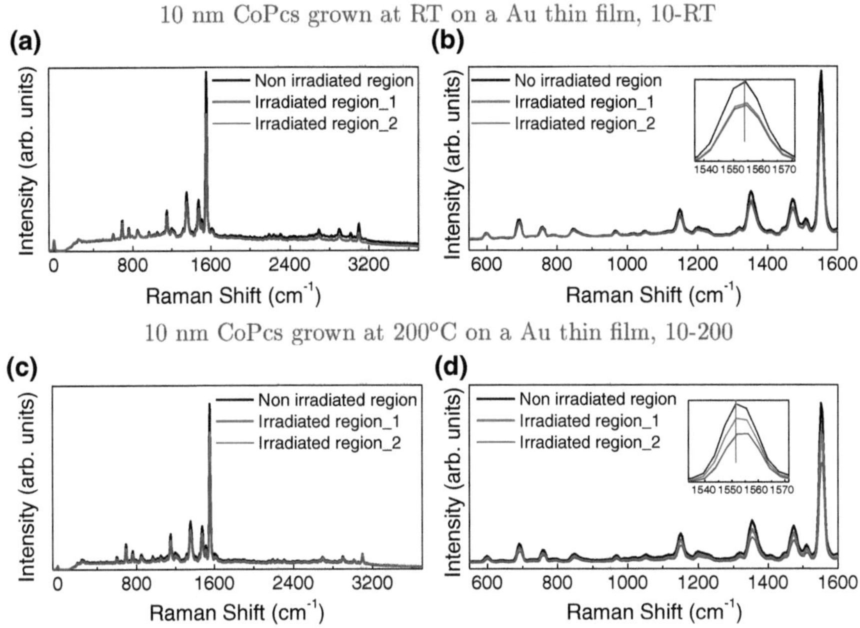

Fig. 7.13 Raman spectra on irradiated and non-irradiated regions for **a** 10-RT and **c** 10–200. **b** and **d** a detail from 550 to 1600 cm^{-1} of Raman spectra

5-RT (see Fig. 7.14), by the optical micrograph (see Fig. 7.14a), it is possible to observe that CoPc layer is quite homogeneous. However, slight intensity changes are observed by in-plane Raman intensity image (Fig. 7.14b), which was obtained from mapping different single Raman spectra measured each 200 nm and integrating in the spectral range from 0 to 3600 cm^{-1}. These modifications (Fig. 7.14c, d) are compared with those observed on the irradiated regions of the CoPc/Au bilayers (see Table 7.3) and we observe that variations of Raman modes are the same order of magnitude for regions of sample 5-RT without irradiating than those between irradiated regions and without irradiating and between two irradiated regions of samples 2-RT and 10–200. Raman measurements were performed with a spot size of around 360 nm (using an objective with a lens of numerical aperture (NA) 0.9). Thus, the variations of Raman intensity can be associated with different local concentrations of CoPc layer.

We cannot discard any of the last two possible origins for the differences observed in the Raman spectra. Consequently, we cannot conclude from the Raman study that X-ray irradiation induces any type of structural modification in the CoPc layers.

By SPR measurements, changes around 1.5–5 % of the refractive index for CoPc layers due to the X-ray irradiation were observed, but detectable modifications are not observed in the CoPc structure by Raman spectroscopy. Thus, Raman measurements

Table 7.3 Parameters correspond to lorentzian fits of four Raman bands of one non-irradiated region and two irradiated regions, for samples 2-RT and 10–200 and for three regions without irradiating of sample 5-RT

	2 nm CoPc layer grown at RT on a Au thin film, 2-RT			
	Raman modes	ν (cm^{-1})	FWHM (cm^{-1})	Intensity (arb. units)
Non-irradiated region	$A_{1g}(5)$	1150	16	8
	$A_{1g}(8)$	1354	23	14
	$B_{1g}(9)$	1474	21	11
	$B_{1g}(10)$	1554	15	61
Irradiated region 1	$A_{1g}(5)$	1150	16	10
	$A_{1g}(8)$	1354	22	18
	$B_{1g}(9)$	1474	20	12
	$B_{1g}(10)$	1553	14	77
Irradiated region 2	$A_{1g}(5)$	1150	15	12
	$A_{1g}(8)$	1354	24	22
	$B_{1g}(9)$	1474	21	13
	$B_{1g}(10)$	1553	14	97
	10 nm CoPc layer grown at 200 °C on a Au thin film, 10–200			
	Raman modes	ν (cm^{-1})	FWHM (cm^{-1})	Intensity (arb. units)
Non-irradiated region	$A_{1g}(5)$	1150	17	673
	$A_{1g}(8)$	1353	27	1070
	$B_{1g}(9)$	1473	18	894
	$B_{1g}(10)$	1554	16	4406
Irradiated region 1	$A_{1g}(5)$	1150	16	433
	$A_{1g}(8)$	1353	23	690
	$B_{1g}(9)$	1473	20	699
	$B_{1g}(10)$	1554	14	2915
Irradiated region 2	$A_{1g}(5)$	1150	16	541
	$A_{1g}(8)$	1353	25	917
	$B_{1g}(9)$	1473	21	839
	$B_{1g}(10)$	1554	15	3629
	5 nm CoPc layer grown at RT on a Au thin film, 5-RT			
	Raman modes	ν (cm^{-1})	FWHM (cm^{-1})	Intensity (arb. units)
Region 1	$A_{1g}(5)$	1150	14	221
	$A_{1g}(8)$	1352	23	366
	$B_{1g}(9)$	1472	24	287
	$B_{1g}(10)$	1551	11	1681

(continued)

Table 7.3 (continued)

	5 nm CoPc layer grown at RT on a Au thin film, 5-RT			
	Raman modes	ν (cm^{-1})	FWHM (cm^{-1})	Intensity (arb. units)
Region 2	$A_{1g}(5)$	1150	15	271
	$A_{1g}(8)$	1352	24	481
	$B_{1g}(9)$	1472	24	371
	$B_{1g}(10)$	1551	12	2124
Region 3	$A_{1g}(5)$	1150	16	214
	$A_{1g}(8)$	1352	24	368
	$B_{1g}(9)$	1472	15	290
	$B_{1g}(10)$	1551	12	1693

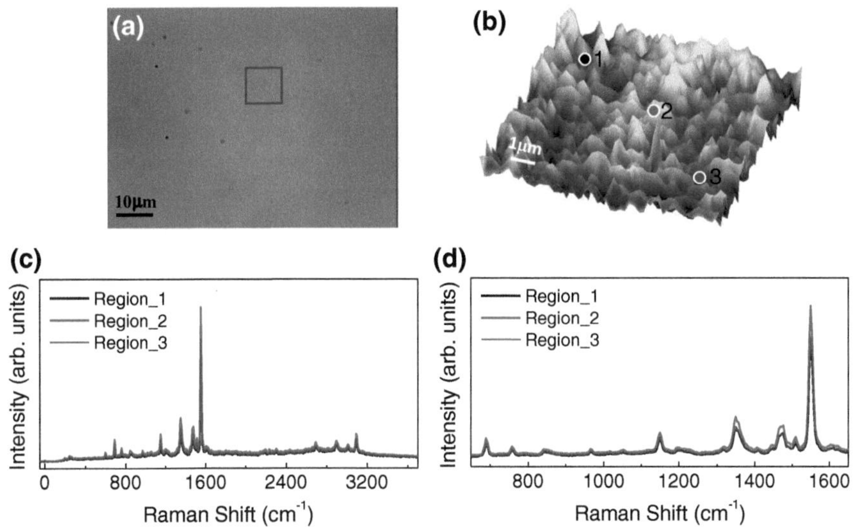

Fig. 7.14 Raman characterization for sample 5-RT: **a** optical micrograph showing the mapped area with a *purple square* (image on plane), **b** in-plane Raman intensity image obtained from mapping the different single Raman spectra taken each 200 nm and integrating in the spectral range from 0 to 3600 cm^{-1}, **c** three single Raman spectra of in-plane image taken from the three region marked with *circles* and **d** a detail of the spectra from 650 to 1650 cm^{-1}. Measurements were performed using a spot size of around 360 nm

of samples were not decisive to assign a correlation between the SPR variations with any change in the Raman spectra. We cannot conclude that X-ray irradiation does not induce structural changes but just that those changes, if exist, are below our experimental sensitivity or they were recovered in the interim between irradiation and Raman measurements (2 months after irradiation).

As mentioned above, Cook et al. observed that the irradiation with soft X-rays on Pc films modifies the Fermi level [12]. This phenomenology is analogous to other semiconductors when introducing a very small number of gap states by dopants, deep traps, defects or interface states [32]. The refractive index of a material can be modified by changes in the lattice structure or through electronic modifications [33]. Structural modifications were not found using Raman spectroscopy. Thus, variations of the Fermi level between the HOMO and LUMO might induce modifications in the optical properties of a material [34]. According to Cook et al., the more subtle radiation effect creates a small minority of defect states, which are insignificant for the chemical composition but they vary the Fermi level position. This Fermi level shift increases with the irradiation time until saturating. With our device, we observed experimentally modifications of SPR spectra due the irradiation with hard X-rays, which are related to changes on the refractive index of CoPc layers and we observed a slow recovering of them. So, these variations may be explained by a modification of the Fermi level. Moreover, this explanation would be consistent with absence of structural changes in the Raman study. We found that the SPR spectra change with the sample thickness and growth temperature. The changes of SPR reflectivity and the FWHM are larger for samples grown at RT than for those grown at 200 °C and independently of the growth temperature, thicker CoPc films show larger FWHM modifications of SPR spectra. However, the refractive index changes, and hence the possible modifications induced in the Fermi level by X-ray irradiation, do not show a tendency neither with CoPc thickness nor with growth temperature. The observed effects in the Pcs due to the X-ray irradiation are quite subtle and according to Cook et al. they might be faded by annealing. However, it is necessary to take into account them for the design of organic device based on Pcs.

7.6 Conclusions

We have studied the effect of hard X-ray irradiation, around Co K-edge, on CoPc layers grown at RT and 200 °C onto Au film/silica substrate system. For this aim, we have used the recently developed experimental setup described in Chap. 5, capable of simultaneously measuring XAS and SPR at a synchrotron beamline. Using SPR spectroscopy as a probe, we could monitor the subtle induced changes of the refractive index for CoPc layers by X-ray irradiation, which are around 1.5–5 %, depending on layer thickness and growth temperature. Thus, CoPcs can be considered robust dye molecules against the hard X-ray irradiation. Monitoring, in real time and in situ, the reflectivity as a function of time, we determined that the induced effects on samples are accumulative with the time and slowly recovered. We could not resolve any structural modification of CoPcs by X-ray irradiation, however changes of refractive index induced by X-rays on CoPcs may be associated with electronic modifications.

References

1. G.L. Pakhomov, Conductivity measurements on phthalocyanine films: evaporated vs. pressure contacts. Solid State Commun. **134**(7), 491–495 (2005)
2. C.C. Leznoff, A.B.P. Lever (eds.), *Phthalocyanines: Properties and Applications*, vol. 1–4 (VCH, Weinheim, 1989)
3. R. Service, Organic leds look forward to a bright, white future. Science **310**(5755), 1762–1763 (2005)
4. P. Singh, N. Ravindra, Optical properties of metal phthalocyanines. J. Mater. Sci. **45**(15), 4013–4020 (2010)
5. A.N. Fernandes, T.H. Richardson, Conductometric gas sensing studies of tert-butyl silicon-[bis ethyloxy]-phthalocyanine LB films. J. Mater. Sci. **43**(4), 1305–1310 (2008)
6. J. Xue, S. Uchida, B.P. Rand, S.R. Forrest, 4.2 % efficient organic photovoltaic cells with low series resistances. Appl. Phys. Lett. **84**(16), 3013–3015 (2004)
7. S. Heutz, C. Mitra, W. Wu, A. Fisher, A. Kerridge, M. Stoneham, T. Harker, J. Gardener, H.H. Tseng, T. Jones, C. Renner, G. Aeppli, Molecular thin films: a new type of magnetic switch. Adv. Mater. **19**(21), 3618–3622 (2007)
8. N.B. McKeown (ed.), *Phthalocyanine Materials: Synthesis, Structure, and Function* (Cambridge University Press, Cambridge, 1998)
9. F. Petraki, H. Peisert, I. Biswas, T. Chassé, Electronic structure of co-phthalocyanine on gold investigated by photoexcited electron spectroscopies: indication of Co ion-metal interaction. J. Phys. Chem. C **114**(41), 17638–17643 (2010)
10. S. Bhattacharjee, B. Brena, R. Banerjee, H. Wende, O. Eriksson, B. Sanyal, Electronic structure of Co-phthalocyanine calculated by GGA+U and hybrid functional methods. Chem. Phys. **377**, 96–99 (2010)
11. G. De La Torre, C. Claessens, T. Torres, Phthalocyanines: old dyes, new materials. putting color in nanotechnology. Chem. Commun. **20**, 2000–2015 (2007)
12. P.L. Cook, P.S. Johnson, X. Liu, A.L. Chin, F.J. Himpsel, Radiation damage in biomimetic dye molecules for solar cells. J. Chem. Phys. **131**(21), 214702 (2009)
13. G. Cabailh, J. Wells, I. McGovern, A. Vearey-Roberts, A. Bushell, D. Evans, Synchrotron radiation studies of the growth and beam damage of tin-phthalocyanine on GaAs (0 0 1)-1 *x* 6 substrates. Appl. Surf. Sci. **234**(1–4), 144–148 (2004)
14. B. Joseph, C. Menon, Studies on the optical properties and surface morphology of cobalt phthalocyanine thin films. E J. Chem. **5**(1), 86–92 (2008)
15. C. Miller, A. Sharoni, G. Liu, C. Colesniuc, B. Fruhberger, I. Schuller, Quantitative structural analysis of organic thin films: an x-ray diffraction study. Phys. Rev. B **72**(10), 1–6 (2005)
16. K.P. Gentry, T. Gredig, Asymmetric grain distribution in phthalocyanine thin films. Phys. Rev. B **80**(17), 7–11 (2009)
17. S. Ikeda, H. Yamakawa, M. Kiguchi, M. Nakayama, K. Saiki, T. Shimada, T. Miyadera, K. Tsutsui, Y. Wada, Thickness dependent characteristics of a copper phthalocyanine thin-film transistor investigated by in situ FET measurement system. Mol. Cryst. Liquid Cryst. **455**(1), 347–351 (2006)
18. A. Serrano, C. Monton, I. Valmiansk, F. Gálvez, A.L. Cortajarena, J. de la Venta, I.K. Shuller, M.A. García, Study of Co-phthalocyanine films by surface plasmon resonance spectroscopy, J. Appl. Phys. **115**, 103–106 (2014)
19. W. Knoll, Interfaces and thin films as seen by bound electromagnetic waves. Annu. Rev. Phys. Chem. **49**, 569–638 (1998)
20. A. Serrano, O. Rodríguez de la Fuente, V. Collado, J. Rubio-Zuazo, C. Monton, G. Castro, M. García, Simultaneous surface plasmon resonance and x-ray absorption spectroscopy. Rev. Sci. Instr. **83**(8), 083101 (2012)
21. A. Serrano, F. Galvez, O. Rodríguez de la Fuente, M.A. Garcia, X-ray irradiation of soda-lime glasses studied in situ with surface plasmon resonance spectroscopy. J. Appl. Phys. **113**(11), 113104 (2013)

22. See http://www.mpip-mainz.mpg.de/knoll/soft/
23. N. Saliba, D.H. Parker, B.E. Koe, Adsorption of oxygen on Au (111) by exposure to ozone. Surf. Sci. **410**(2–3), 270–282 (1998)
24. T. Jawhari, Micro-raman spectroscopy of the solid state: applications to semiconductors and thin films. Analusis **28**(1), 15–21 (2000)
25. J. Park, J.E. Royer, C.N. Colesniuc, F.I. Bohrer, A. Sharoni, S. Jin, I.K. Schuller, W.C. Trogler, A.C. Kummel, Ambient induced degradation and chemically activated recovery in copper phthalocyanine thin film transistors. J. Appl. Phys. **106**(3), 034505 (2009)
26. D.R. Tackley, G. Dent, W. Ewen Smith, Phthalocyanines: structure and vibrations. Phys. Chem. Chem. Phys. **3**, 1419–1426 (2001)
27. M. Szybowicz, W. Bala, S. Dümecke, K. Fabisiak, K. Paprocki, M. Drozdowski, Temperature and orientation study of cobalt phthalocyanine CoPc thin films deposited on silicon substrate as studied by micro-Raman scattering spectroscopy. Thin Solid Films **520**(1), 623–627 (2011)
28. M. Szybowicz, W. Bala, K. Fabisiak, K. Paprocki, M. Drozdowski, Micro-Raman spectroscopic investigations of cobalt phthalocyanine thin films deposited on quartz and diamond substrates. Cryst. Res. Technol. **45**(12), 1265–1271 (2010)
29. M.-S. Liao, S. Scheiner, Electronic structure and bonding in metal phthalocyanines, metal=Fe Co, Ni, Cu, Zn, Mg. J. Chem. Phys. **114**(22), 9780 (2001)
30. B.J. Palys, D.M.W. van den Ham, W. Briels, D. Feil, Resonance Raman spectra of phthalocyanine monolayers on different supports. A normal mode analysis of zinc phthalocyanine by means of the MNDO method. J. Raman Spectr. **26**, 63–76 (1995)
31. M. Fleischmann, P.J. Hendra, A.J. Mcquillan, Raman spectra of pyridine adsorbed at a silver electrode. Chem. Phys. Lett. **26**(2), 163–166 (1974)
32. M.S. Dresselhaus, *Solid State Physics. Part II: Optical Properties of Solids*. MIT Solid State Physics Course, 2001
33. C. Kittle, *Introduction to Solid State Physics*, 8th edn. (Wiley, New York, 2005)
34. P. Hervé, L. Vandamme, General relation between refractive index and energy gap in semiconductors. Infrared Phys. Technol. **35**(4), 609–615 (1994)

Chapter 8
Conclusions

According to the main results obtained in this work, the most relevant conclusions are presented below:

Study by Surface Plasmon Resonance Spectroscopy of Au Films and Au/Fe Bilayers Annealed Under Different Conditions

In this first part of this work, Au films and Au/Fe bilayers, annealed under different conditions, have been studied with surface plasmons (SPs). Parameters such as the film thickness and the annealing conditions have been modified.

- With regards to the Au films annealed in air at different temperatures, the following conclusions can be inferred:
  - Annealing of Au thin films deposited on soda-lime glass substrates promotes the transition from films to Au islands. The features of the obtained particles, such as size, shape, height and inter-island distance depend on both Au film initial thickness and annealing temperature.
  - The structural modifications induced in the samples changing the film thickness and the annealing temperature, lead to significant variations in the SPs. With the annealing temperature, we have followed the progressive change from extended SPs (ESPs) to localized SPs (LSPs). Upon annealing, films with thickness up to 30 nm exhibit an optical absorption associated with LSPs, while thicker films can exhibit ESPs (Au islands show dimensions larger than the wavelength visible of light).
- With regards to the Au/Fe bilayers annealed under different conditions in air or low vacuum, we can conclude that:

A. Serrano Rubio, *Modified Au-Based Nanomaterials Studied by Surface Plasmon Resonance Spectroscopy*, Springer Theses, DOI 10.1007/978-3-319-19402-8_8

– Annealing, under specific conditions, of Au/Fe bilayers grown on silica substrates shows a new method to fabricate Au/FeO_x nano and microparticles. The morphological features of the complex particles depend on parameters such as Fe initial thickness (Au initial thickness has not varied), the disposition of the layers and the annealing conditions (i.e., temperature, atmosphere). The method have been successfully applied to the fabrication of Au/α-Fe_2O_3 and Au/α-Fe_2O_3-γ-Fe_2O_3 nanostructures ensembles, although this method could be applied to a wide range of composed materials.
– The different strategies carried out for the fabrication of the complex nanoparticles have allowed growing separately Au and FeO_x nanostructures or complex dimers, using the proper annealing conditions. Au and FeO_x particles have been identified by confocal Raman microscopy. The diverse morphological properties of the nanostructures have led to variations in the surface plasmon resonance (SPR) and its profiles have been associated with the different structures. Besides the observation of SPR in complex nanostructures, these may show additional functionalities.

Therefore, annealing of Au films or Au/Fe bilayers deposited on glass substrates provides a method to tune the morphological and plasmonic properties in a wide range and over large areas not easily achievable by other methods.

Design and Development of the SPR-XAS Setup. Study by Surface Plasmon Resonance Spectroscopy of the Interaction of X-rays with Matter

In this second set of experiments, modifications induced in matter by an external source, in real time and *in situ*, have been carried out using SPR as probe, in order to study the optical changes. For that purpose, we have designed and developed a setup combining SPR and X-ray absorption spectroscopy: the SPR-XAS setup, which is currently available for future experiments at the ESRF beamline BM25A-SpLine. With this device, SPR is used to study the refractive index changes induced by X-rays, specifically, on glasses and Co-phthalocyanines (CoPcs). The system versatility allows measuring simultaneously SPR excitation and X-ray absorption while scanning different parameters, with a resolution of the measurements of the order of 10^{-2} to 10^{-5}, depending on the particular experiment.

- With respect to the study by SPR of glasses irradiated with hard X-rays, Au films are grown on glass substrates using two different glass substrates: soda-lime and silica substrates. From this study, the following conclusions are inferred:

  – For soda-lime glasses, a darkening process has been observed, which is associated with two types of defects: HC1 and HC2 centers. These are responsible for changes in both the real and the imaginary parts of the refractive index of

soda-lime glasses upon X-ray irradiation. Measuring the reflectivity versus time, two different dynamics by recombination with electrons in both the formation and healing of the induced defects are determined. Besides this, slow and fast processes are identified with characteristic times of the order of months and minutes, respectively. The fast component has not been previously reported and has only been detectable thanks to our experimental setup, measuring *in situ*.
- For the case of silica glasses, we have not identified color centers that modify their refractive index. However, slight variations in the experimental SPR spectra have been observed and attributed to changes in where the Au film is involved.

- With respect to the study by SPs of CoPc films irradiated with hard X-rays, CoPc/Au samples are grown on silica substrates varying the CoPc thickness and growth conditions. From that, we can conclude that:
  - Using SPR as a probe, SPR-XAS setup has allowed us to study the refractive index modifications induced with X-rays on the CoPcs, which are around 1.5–5 %, depending on CoPc film thickness and growth temperature. Measuring the reflectivity versus the time, we have determined that the induced effects by X-rays on CoPcs are accumulative with the time and slowly recover.
  - We have not found any structural modification induced by X-rays in the CoPc films. However, we have observed changes of their refractive index that may be associated with electronic modifications induced by the irradiation with X-rays.

Appendix A
Estimation of Changes in the SPR Curve upon X-ray Irradiation

In order to estimate the expected changes in the SPR curve of a Au film upon X-ray illumination, we calculate initially the change in conduction electrons in our material upon irradiation and the variation that it induces in the SPR curve.

The density of conduction electrons for Au is 5.9×10^{22} electrons/cm^3 [1]. The irradiated region of the sample has an area of ~1 mm^2 and a thickness of 50 nm. Therefore, the number of conduction electrons in this region in absence of X-ray illumination is $n_c = 2.95 \times 10^{15}$.

The Au absorption coefficient for X-rays with energy of 11.9 keV is 0.3 μm^{-1} [2]. Therefore, if the incident beam has N_0 photons, according to the Lambert-Beer law at 50 nm the number of photons in the beam is

$$N(50\,\text{nm}) = N_0 e^{(-0.3\ \mu\text{m}^{-1} \times 50\ \text{nm})} \tag{A.1}$$

Consequently, the number of absorbed photons in the film is

$$N_{abs} = N_0(1 - e^{(-0.3\ \mu\text{m}^{-1} \times 0.05\ \mu\text{m})}) \approx N_0(1 - 1 + 0.015) \approx 0.015 N_0 \tag{A.2}$$

The photon flux of the incident beam is about $N_0 = 10^{12}$ photons/s and therefore the photons absorbed in the irradiated region of the film is $N_{abs} \approx 1.5 \times 10^{10}$ photons/s.

The change in photoexcited conduction electrons, n, is given by:

$$\frac{dn}{dt} = 1.5 \times 10^{10}\ \text{s}^{-1} - \frac{n}{\tau} \tag{A.3}$$

In the steady state upon continuous illumination $\frac{dn}{dt} = 0$ and therefore $n = 1.5 \times 10^{10} \times \tau\ \text{s}^{-1}$.

The lifetime of photoexcited electrons in metals is of the order of femtoseconds. However, the relaxation process takes place commonly by X-ray fluorescence and Auger effect [3, 4] where electrons are excited from inner levels to above the Fermi

A. Serrano Rubio, *Modified Au-Based Nanomaterials Studied by Surface Plasmon Resonance Spectroscopy*, Springer Theses, DOI 10.1007/978-3-319-19402-8 level, increasing the effective time in which we have an excess of population at the conduction band. A proper estimation of the time in which we have excited electron in the conduction band is given by the decay time of X-ray fluorescence excited by X-rays. For metals these times are of the order of 10–100 ns. Thus, this is the effective decay time that we consider for our estimation. With this number, in the steady state the number of photoexcited electrons is of the order of 10^3. Consequently, the relative change of conduction electrons in the Au film is: $\frac{\Delta n}{n} \approx 10^{-12}$.

The plasma frequency of a metal is given by

$$\omega_p = \sqrt{\frac{4\pi n e^2}{m}} \tag{A.4}$$

being n the density of conduction electrons and e and m the electron charge and mass, respectively. Thus, the relative change in plasma frequency induced by a change in conduction electrons is $\frac{\Delta\omega_p}{\omega_p} = \frac{\Delta n}{2n} \approx 5 \times 10^{-13}$.

Within the Drude model, the dielectric permittivity of a metal is given by

$$\varepsilon = 1 - \frac{\omega_p^2}{\omega^2} \tag{A.5}$$

The relative change in dielectric permittivity induced by a modification of the plasma frequency is given by

$$\frac{\Delta\varepsilon}{\varepsilon} = \frac{2\frac{\omega_p}{\omega^2}}{1 - \frac{\omega_p^2}{\omega^2}} \Delta\omega_p = \frac{2\frac{\omega_p^2}{\omega^2}}{1 - \frac{\omega_p^2}{\omega^2}} \frac{\Delta\omega_p}{\omega_p} \tag{A.6}$$

The plasma frequency for gold [5] is $\omega_p = 2.18 \times 10^{15}\ \mathrm{s}^{-1}$ and in our experiment, we use a red laser with wavelength 632.8 nm and frequency $4.74 \times 10^{14}\ \mathrm{s}^{-1}$. Thus $\frac{\omega_p^2}{\omega^2} \approx 20$ and $\frac{\Delta\varepsilon}{\varepsilon} \approx -2\frac{\Delta\omega_p}{\omega_p} \approx 10^{-12}$.

We will not discuss the particular changes in the real and imaginary part of the dielectric function since we are just providing an estimation.

For the SPR spectrum of a 50 nm Au film, a relative change in the dielectric permittivity of the order of 10^{-3} induces variation in the resonance angle of the order of 10^{-2} and, if we consider the reflected intensity in the region of the curve with higher slope this change is around 0.05. Thus a change in the dielectric permittivity of the order of 10^{-12} can induce changes in the reflected intensity of the SPR curve of $\sim 10^{-10}$.

References

1. C. Kittle, *Introduction to Solid State Physics*, 8th edn. (Wiley, New York, 2005)
2. See http://physics.nist.gov/PhysRefData/XrayMassCoef/ElemTab/z79.html
3. See http://webhost.ua.ac.be/mitac4/micro_xrf.pdf
4. G. Gauglitz, T. Vo-Dinh (eds.), *The Handbook of Spectroscopy (Chapter X-ray analysis)* (Wiley-VCH, New York, 2003)
5. M.A. Ordal, L.L. Long, R.J. Bell, R.S. Bell, R.R. Bell, J.R.W. Alexander, C.A. Ward, Optical properties of the metals Al Co, Cu, Au, Fe, Pb, Ni, Pd, Pt, Ag, Ti, and W in the infrared and far infrared. Appl. Opt. **22**, 1099–1119 (1983)

MIX
Papier aus verantwortungsvollen Quellen
Paper from responsible sources
FSC® C105338
www.fsc.org

If you have any concerns about our products,
you can contact us on
ProductSafety@springernature.com

In case Publisher is established outside the EU,
the EU authorized representative is:
Springer Nature Customer Service Center GmbH
Europaplatz 3, 69115 Heidelberg, Germany

Printed by Libri Plureos GmbH
in Hamburg, Germany